CARPENTRY LAYOUT

by

Ken Todd

CRAFTSMAN BOOK COMPANY
6058 Corte del Cedro, P.O. Box 6500, Carlsbad, CA 92008

To Zoey, Annika and Megan,
for their help and encouragement
in putting this book together

Library of Congress Cataloging in Publication Data

Todd, Ken.
  Carpentry layout / by Ken Todd.
    p. cm.
  Includes index.
  ISBN 0-934041-32-6
  1. Laying-out (Woodwork) 2. Pocket calculators. I. Title.
TH5615.T62 1988   88-4561
694'.2-dc19   CIP

©1988 Craftsman Book Company

Illustrated by Steven Silvera

Contents

1. **How to Choose a Calculator** 5
   How to Use This Book 7

2. **Equally Spaced Courses** 9
   Calculating Courses 9
   Boards and Battens 10
   Decking 11
   Shingles and Shakes 12
   Pickets and Balusters 13
   Paneling or Siding 13
   Beginning Work 14
   Applications 18
   Problems 37

3. **Stair Layout** 38
   Calculating Stair Layout 38
   Terms 39
   Beginning Work 42
   Applications 43
   Problems 53

4. **Stair Layout in the Field** 55
   Laying Out the Stairs 55
   Stair Carriages 57
   Installing the Stairs 60

5. **Common Rafter Layout** 63
   Terms 63
   What You'll Need to Know 64
   Beginning Work 68
   Applications 72
   Problems 96

6. **Rafter Layout in the Field** 99
   Cutting Rafters 99
   The Tail Cut 100
   The Birdsmouth Cut 100
   Fitting the Rafters 101

7. **Bearing-Wall Heights** 103
   Calculating the Heights 103
   Terms 105
   Beginning Work 105
   Applications 106
   Problems 116

8	**Rake-Wall Layout**	117
	Terms	118
	What You Need to Know	120
	Beginning Work	121
	Applications	125
	Problems	145
9	**Calculating Foundation Layouts**	148
	Tools You'll Need	148
	Using Your Calculator	148
	Beginning Work	150
	Applications	152
	Problems	166
10	**Foundation Layout in the Field**	167
	Laying Out the Points	167
11	**Calculating Hip, Valley and Jack Rafters for Equal-Pitch Roofs**	173
	Hip Rafters	174
	Beginning Work	176
	Applications	177
	Valley Rafters	183
	Applications	185
	Jack Rafters	187
	Beginning Work	188
	Applications	190
	Problems	196
12	**Practical Application: Laying Out Hip, Valley and Jack Rafters for Equal-Pitch Roofs**	200
	Laying Out Rafters	200
	Side Cuts	204
	Setting Jack Rafters	210

Appendix A: Answers for Chapter Problems 211

Appendix B: Height, Rake, and Diagonal Multipliers 215

Appendix C: Fraction to Decimal Conversions 216

Appendix D: Decimal to Fraction Conversions 216

Index 217

1
How to Choose a Calculator

Every carpenter knows that there's more to carpentry than just driving a straight nail. In fact, that's probably the smallest part of the job. An experienced carpentry pro is fast, accurate, and innovative. The pro is always finding better, more productive ways to do highly professional work.

But there's even more to learn about carpentry than the physical skills most carpenters have already mastered. Erecting stairs, formwork, rafters, joists and rake walls is the easy part. Figuring the quantities of materials needed and cutting them to exact lengths is the hard part. Work like that is usually left to the most experienced, best-paid carpenter on the job. We'll call this work **carpentry layout** because it involves visualizing how the parts go together, calculating the exact lengths and figuring quantities of materials.

For centuries carpenters used the framing square as a calculator to figure dimensions and angles. It was a good tool in its time. But the modern hand calculator makes the framing square obsolete for most of the layout work carpenters have to do. If you don't believe me, keep reading. Even tough problems are easy when you have the power of a modern microprocessor in your hands. The square will be around for a while. But 20 years from now, only the old-timers will still be using squares to solve layout problems.

If you've avoided layout problems until now, the modern calculator may be what you've been waiting for. Read and use this book and I think you'll agree.

Inevitably, this book involves mathematics. If you were never good at math, don't worry. The calculator does all the figuring for you. I intend to take it slow and easy. You should have no trouble following my explanations. I'm a carpenter and I've taught many carpenters how to use a calculator to solve layout problems. So I know what's hard to grasp and which explanations make sense to someone who's just learning layout with a calculator. Stay with me. I'm going to teach you skills you never expected to master.

You can imagine the difference this valuable skill can make in your work. If you are already doing layout, you can now figure more complicated problems faster and with more confidence. If you have avoided layout, you can now master this important area of knowledge, expanding your skills and increasing your value as a carpenter. This is the inevitable future of carpentry. Just as power saws have replaced hand saws, layout methods involving calculators will replace the traditional ways of figuring layout. Progressive builders need carpenters who can use calculators to do layouts fast and accurately.

Good-quality calculators are available at modest cost in department, drug, and discount stores. There are three basic kinds of logic used in calculators: algebraic, sequential and reverse Polish logic. These terms refer to the way in which the calculator responds to the keys you push. We recommend that you avoid calculators with reverse Polish logic. Algebraic calculators are best for carpentry work. However, if you already have a sequential calculator, you may use it, but you will have to modify the calculator sequences given in this manual. All calculations in this book are done on an algebraic calculator.*

For carpentry work, you need a calculator that has all the following functions: Addition ⊞, subtraction ⊟, multiplication ⊠, division ⊡, square root √x̄, square x^2, tangent tan, polar coordinates P▸R, or R▸P, five or more memories MEM or STO, recall RCL, conversions from degrees, minutes and seconds to decimal degrees DMS-DD, inverse function INV, and programming capability.

Do not be concerned if you don't understand what all these functions are at this point. This book explains clearly how and where to use each function. This book is set up to allow you to use your calculator to do the computations without knowledge of trigonometry or, indeed, any advanced mathematics.

* To determine if a calculator is algebraic or sequential, perform the following test:

$$1 \boxplus 2 \boxtimes 5 \boxminus$$

If the answer on display is 11, you have an algebraic calculator.
If the answer on display is 15, you have a sequential calculator.

If you already have a calculator which will only add ⊞, subtract ⊟, multiply ⊠, divide ⊡, and compute square roots √x̄, you can do all the calculations in this book except rake-wall layout (Chapter 8) and foundation layout (Chapter 9).

As you look at several calculators with algebraic logic and the required functions (there will probably be other functions you won't need on the calculators), consider other features, such as cost and color display and size of display (large numbers are best, and green numbers seem to be easier to read in sunlight than red). Some calculators are also available with a printout (the answer is given on paper as well as in a lighted display), which is not necessary for these purposes but might be useful for other tasks. The size of the buttons is an important consideration, too. You should avoid calculators with buttons too small or too closely spaced. The action or feel of the buttons when pushed should be smooth. You may also need an adaptor, which will recharge specially purchased batteries, and a carrying case (preferably a hard-shell type) to protect your calculator.

How to Use This Book

Each chapter in this book is about one area of construction layout. The first part of the chapter examines the topic in general and tells how to solve layout problems in this area. Then, a step-by-step procedure to use in solving the problems is given. Next, specific problems using the procedure are presented. Each chapter ends with some problems for you to solve along with some worksheets to use in solving the problems. The answers are at the end of the book.

When describing a sequence of buttons to push, we will put the functions in a box. For example, if we want you to push the key to add, we will use this: ⊞. For the key to multiply, we will use ⊠. An example of a button sequence is as follows:

(Enter)	(plus)	(Enter)	(equals)	(display)	(times)	(Enter)	(equals)	(display)
15	⊞	9	⊟	24	⊠	2	⊟	48

When we give a sequence of directions for you to follow on your calculator, boldface words in parentheses indicate quantities that you will replace with numbers when calculating. For example,

(Total run) ⊠ **(diagonal multiplier)** ⊟ **(length of hip rafter)**

means to replace **(total run)** and **(diagonal multiplier)** with numbers. The length of the hip rafter will be displayed.

I used a Texas Instruments TI-55-II calculator to perform the calculations in this book, but this is by no means the only calculator that will work. Virtually every calculator company each produces two or three calculators which would be suitable.

Most calculators have the capacity to round off figures on display. I have chosen not to use this feature, but you may wish to consider it. Sometimes your figures will not agree with mine to the last decimal place; don't worry, the difference is not significant (it usually involves millionths of an inch). For general carpentry work you will need to be accurate to within $\frac{1}{16}$ of an inch.

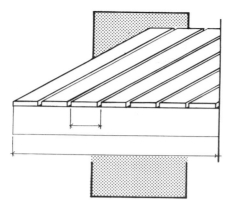

2
Equally Spaced Courses

In this chapter, you'll learn a simple, fast, and accurate way to find the layout of equally spaced courses. A *course* is any repeating unit of materials or materials and spaces. It is encountered in all stages of construction. You figure a course in these situations: joist and rafters when symmetry is desired, placing board-and-batten siding, placing decking material to best advantage, figuring the exposure of shakes and shingles so that the roof looks square, placing paneling symmetrically, and laying out pickets or handrail uprights (balusters).

Calculating Courses

To calculate courses you need to know the following:
1) The *length* of the area the courses are covering, in decimal inches
2) The *largest acceptable course,* in decimal inches

Joists or Rafters
In figuring layout of joists or rafters, the *length* is the distance from the *inside* of the first joist or rafter to the *outside* of the last joist or rafter (Figure 2-1). The *largest acceptable course* is the largest spacing

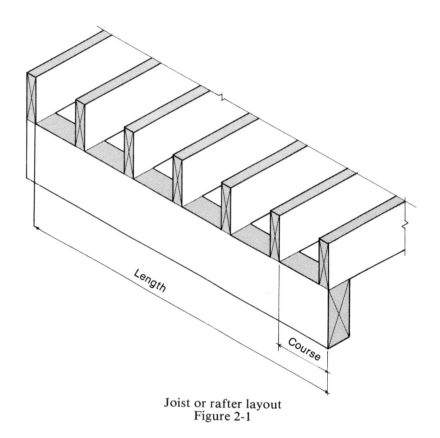

Joist or rafter layout
Figure 2-1

allowable for the joists or rafters. This is measured from the outside of one joist or rafter to the inside of the next joist or rafter. (This is the same as a center-to-center dimension). You may be given this dimension by the architect, engineer, or building code; it is determined by structural necessity or aesthetic considerations. For example, if rafters or rafter tails are exposed, you may want symmetry. For exposed joists, joist ends, or evenly spaced nails on decking, you may want even placement of joists.

Boards and Battens

In figuring board-and-batten spacing, there are two common patterns. In pattern 1 (Figure 2-2), a batten begins the pattern and a board ends it. In this case the *length* is the total length of the wall to be covered. The *largest acceptable course* is the measured board width

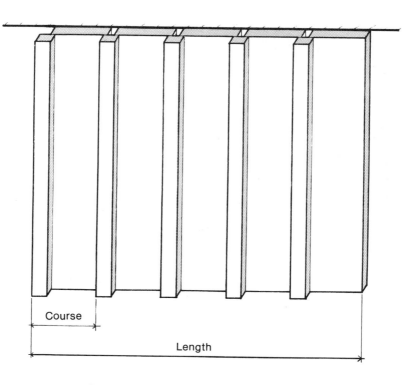

Board-and-batten spacing, pattern 1
Figure 2-2

plus the measured batten width minus the overlap, the width of the board overlapped on each end by the batten.

In pattern 2 (Figure 2-3), a batten begins *and* ends the pattern, so there is a batten at each end. In this case the *length* is the total length of the area to be covered minus the width of one batten. The *largest acceptable course* is the same as in the first case—the measured board width plus the measured batten width minus the overlap.

Decking

For decking, the *length* is the measured length of the joists to be covered. The *largest acceptable course* is the measured width of the decking material plus the largest acceptable space between the boards (Figure 2-4). The largest acceptable space will vary with different decking applications.

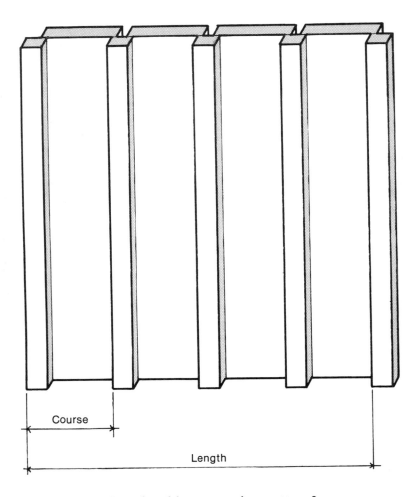

Board-and-batten spacing, pattern 2
Figure 2-3

Shingles and Shakes

In figuring courses of shingles and shakes, the *length* is the distance from the lower edge of the ridgecap course to the lower edge of the eaves plus the desired overhang of the shingles at the eaves. The *largest acceptable course* is the largest exposure of shingles or shakes acceptable, which is determined by the material and use (Figure 2-5). Only an architect or an engineer should change the largest acceptable course.

Equally Spaced Courses   13

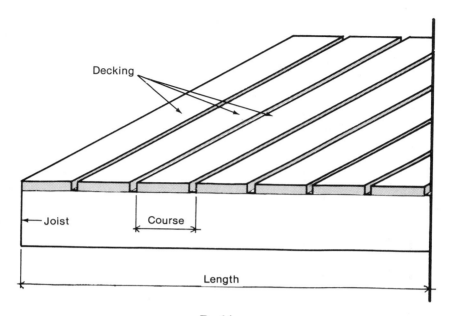

Decking
Figure 2-4

Pickets and Balusters
Pickets and balusters can be arranged in two patterns. In pattern 1 (Figure 2-6), there is a space next to each post. In this case, the *length* is the distance between the inner edges of the two posts *plus* the width of one picket. In pattern 2, pickets are placed immediately adjacent to the post (Figure 2-7). The *length* is the distance between the inner edges of the two posts *minus* the width of one picket. The *largest acceptable course* is the measured width of the picket plus the space next to it. The architect, engineer, or building code may give this measurement.

Paneling or Siding
For paneling or siding, the *length* is the length of the area to be covered. The *largest acceptable course* is the measured width of the material to be used. (Exclude the tongue on tongue-and-groove or shiplap siding.) Since we cannot adjust the width of the material in this case, we want to arrange the panels in a symmetrical fashion to produce a professional job. This will give panels of equal width at each end of the wall (Figure 2-8).

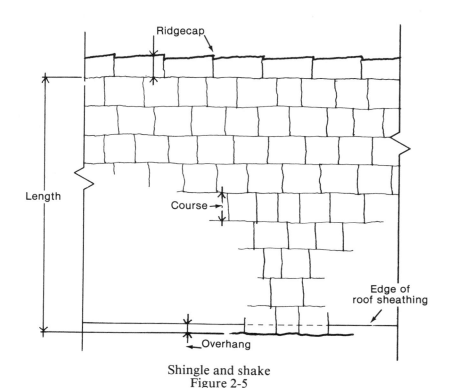

Shingle and shake
Figure 2-5

Beginning Work

You cannot use a calculator directly to work with fractions. When using a calculator, you must therefore convert all fractional measurements to decimal fractions. To convert a measurement in feet and inches to inches, multiply the number of feet by 12. Next add the inches. If there is a fractional number of inches, divide the numerator (top number) by the denominator (bottom number). Add this decimal fraction to the total number of inches.

To convert measurements to decimal inches on a calculator, do the following:

    (Ft.) ⊠ 12 ⊞ **(in.)** ⊞ **(numerator)** ⊟ **(denominator)** ⊟ **(decimal in.)**

Example   Convert 21′ 11⅝″ to decimal inches.

    (Ft.) ⊠ 12 ⊞ **(in.)** ⊞ **(numerator)** ⊟ **(denominator)** ⊟ **(decimal in.)**
     21 ⊠ 12 ⊞ 11 ⊞   5   ⊟   8   ⊟ 263.625

Now that you have the *length* and the *largest acceptable course* in

Equally Spaced Courses **15**

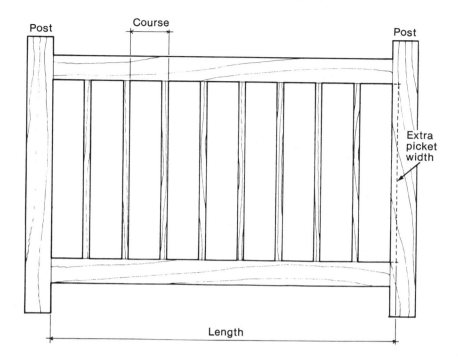

Picket and baluster, pattern 1
Figure 2-6

decimal inches, you are ready to calculate the courses. To do this divide the *length* by the *largest acceptable course* (step 1). Round off this figure to the next greatest whole number. This gives you the *number of courses* to use (step 2). Next, divide the *length* by the *number of courses*. This gives you the exact course dimension in decimal inches (step 3). The last step is to convert the answer back to inches and sixteenths of inches. You do this by multiplying the decimal part of the course dimension by 16 (steps 5 and 6).

Use these steps to find the course dimension on a calculator.
1) **(Length)** ÷ **(largest acceptable course)** = **(display)**
2) Round off the number on display to the next highest whole number; this is the *number of courses* to use.
3) **(Length)** ÷ **(number of courses)** = **(course dimension)**
4) Write down the whole-number part of the course dimension from step 3.

16 *Carpentry Layout*

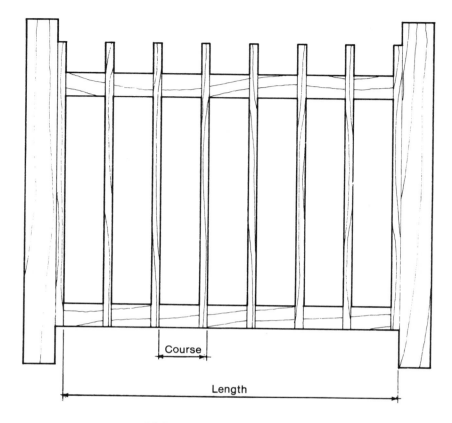

Picket and baluster, pattern 2
Figure 2-7

5) **(Course dimension)** ⊟ **(whole-number part)** ⊟ ⊠ **16** ⊟ **(no. of 16ths)**
6) Round off the number of sixteenths to the nearest whole number. Write a fraction using this number as numerator and 16 as denominator. Add this fraction to the whole number from step 4 to obtain the course dimension in inches and sixteenths of inches.

Example The *length* is 452.375″. The *largest acceptable course* is 32″. Find the exact course dimension.
1) **(Length)** ⊞ **(largest acceptable course)** ⊟ **(display)**
   452.375 ⊞   32   ⊟ 14.136718
2) Round 14.136719 to 15, the next greatest whole number.
3) **(Length)** ⊞ **(number of courses)** ⊟ **(course dimension)**
   452.375 ⊞   15   ⊟   30.158333

Equally Spaced Courses **17**

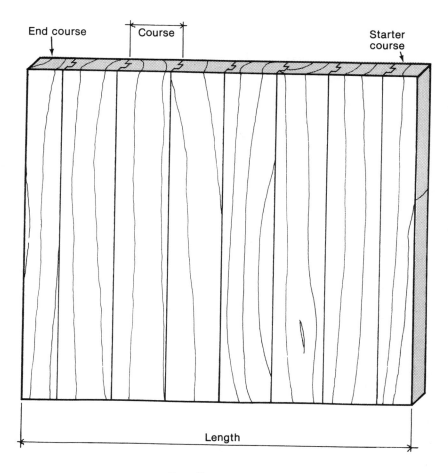

Paneling or siding
Figure 2-8

4) Write down 30, the whole-number part of the course dimension.
5) **(Course dimension)** ⊟ **(whole-number part)** ⊟ ⊠ **16** ⊟ **(no. of 16ths)**
    30.158333   ⊟   30   ⊟ ⊠ 16 ⊟   2.533328
(Note that you do not have to enter 30.158333—it is already displayed.)
6) Round off this number to the nearest whole number to get 3. Write 3/16. Add this to 30 (from step 4). The course dimension is 30 $\frac{3}{16}$″.

You may feel this looks like a lot of complicated steps. However,

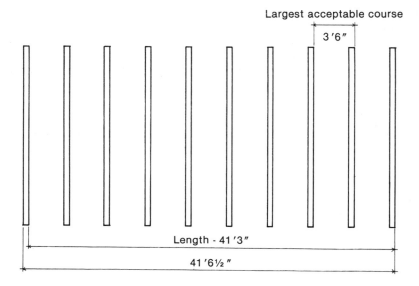

Rafter layout
Figure 2-9

as with any new tool, you will find it easier to use with practice. You will soon find that these procedures are second nature to you, and you can calculate courses faster and more accurately than before. Remember that you are now doing on a calculator what you used to do painstakingly by hand. The following examples illustrate solving specific problems. For best results work through these problems using your own calculator. In the next section, there are problems for you to work on your own.

Applications

Example 1 Rafters
You are working on a house with an exposed rafter ceiling. The rafters are 4″ × 12″ and may not exceed 3′6″ on center. The architect wants them to be equally spaced over the length of the house, which is 41′6½″. What is the course dimension (Figure 2-9)?

A) Find the length in decimal inches.
To find the *length*, the distance from the inside of the first rafter to the outside of the last rafter, subtract the actual thickness of a rafter, 3½″, from the length of the house, 41′6½″. This leaves a *length* of 41′3″. Convert this to decimal inches:

Equally Spaced Courses 19

(Ft.) × 12 + **(in.)** = **(decimal in.)**
41 × 12 + 3 = 495

B) The *length,* in inches, is 495. The *largest acceptable course* is 3'6", which is given by the architect. Convert this to decimal inches:
(Ft.) × 12 + **(in.)** = **(decimal in.)**
3 × 12 + 6 = 42
This is the *largest acceptable course* in inches. Again, there is no fraction to convert.

C) Now that we have the *length* and the *largest acceptable course* in inches, we are ready to calculate the exact course dimension.
   1) **(Length)** ÷ **(largest acceptable course)** = **(display)**
      495 ÷ 42 = 11.785714
   2) Round 11.785714 up to the next greatest whole number, 12. This is the *number of courses* to use.
   3) **(Length)** ÷ **(number of courses)** = **(course dimension)**
      495 ÷ 12 = 41.25
      This is the *course dimension* in decimal inches.
   4) Write down 41".
   5) **(Course dimension)** − **(whole-number part)** = × 16 = **(no. of 16ths)**
      41.25 − 41 = × 16 = 4
   6) Add $\tfrac{4}{16}$" or ¼ " to 41 " (from step 4). The course dimension is 41¼ ".

Example 2 Joists
You are working on a second-floor sun deck. The joists will be exposed to view below, and you want them to be equally spaced. Structural analysis calls for 6 × 10 joists at 4' on center. If the deck is 37' 6⅜" long, what should the course dimension be (Figure 2-10)?

A) First, find the *length.* The deck is 37' 6⅜". Next, convert this to decimal inches:
(Ft.) × 12 + **(in.)** + **(numerator)** ÷ **(denominator)** = **(decimal in.)**
37 × 12 + 6 + 3 ÷ 8 = 450.375
This is the length of the deck in decimal inches.
To find the *length* from the inside of the first joist to the outside of the last joist, subtract the actual width of one joist—5½ ", or 5.5"—from the total length of the deck, 450.375":
450.375 − 5.5 = 444.875
This is the length from the inside of the first joist to the outside of the last joist in decimal inches.

20 *Carpentry Layout*

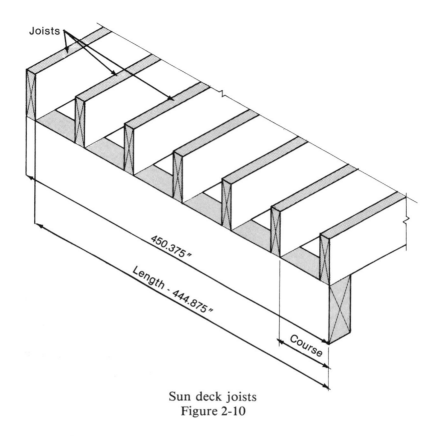

Sun deck joists
Figure 2-10

B) Next, find the *largest acceptable course* in decimal inches. The specified course is 4′. To convert to decimal inches:
4 ⊠ 12 ⊟ 48
This is the *largest acceptable course* in decimal inches.

C) Now, calculate the course dimension.
  1) **(Length)** ⊟ **(largest acceptable course)** ⊟ **(display)**
      444.875 ⊟   48   ⊟ 9.2682292
  2) Round 9.2682292 to the next greatest whole number, 10. This is the *number of courses*.
  3) **(Length)** ⊟ **(number of courses)** ⊟ **(course dimension)**
      444.875 ⊟   10   ⊟ 44.4875
      This is the *course dimension* in decimal inches.
  4) Write down 44″.
  5) **(Course dimension)** ⊟ **(whole-number part)** ⊟ ⊠ 16 ⊟ **(no. of 16^{ths})**
        44.4875   ⊟   44   ⊟ ⊠ 16 ⊟   7.8

Equally Spaced Courses **21**

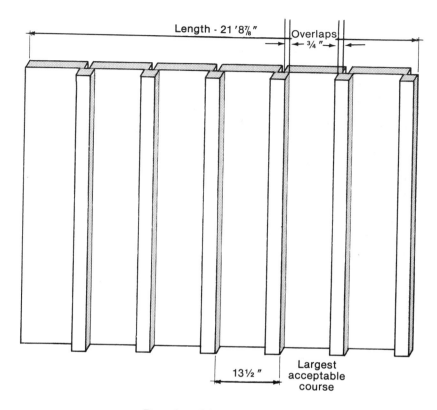

Board-and-batten siding
Figure 2-11

6) Round off 7.8 to the nearest whole number: 7.8 rounds off to 8. Add $8/16''$, or $1/2''$, to 44 (from step 4). The course dimension is $44 1/2''$.

Example 3   Board-and-battens
You are siding a structure using $1'' \times 12''$ boards and $1'' \times 4''$ battens. The battens must overlap each board by at least $3/4''$ for nailing. The siding will start with a board and end with a batten (see Figure 2-11). The wall you are siding is $21' 8 7/8''$ long. If you want all the boards and battens equally spaced with no narrow course at the end, what will be the course dimension?

A) First, find the length of the wall, $21' 8 7/8''$, in decimal inches:

(Ft.)	✕	12	+	(in.)	+	(numerator)	÷	(denominator)	=	(decimal in.)
21	✕	12	+	8	+	7	÷	8	=	260.875

This is the *length* of the wall in decimal inches.

B) Next, find the *largest acceptable course*. This is the actual width of the board, 11½″, plus the actual width of the batten, 3½″, minus the overlaps for nailing, or 1½″:
11.5 ⊞ 3.5 ⊟ 1.5 ⊟ 13.5
This is the *largest acceptable course* in decimal inches.

C) Next, find the course dimension.
   1) **(Length)** ⊡ **(largest acceptable course)** ⊟ **(display)**
      260.875 ⊡   13.5   ⊟ 19.324074
   2) Round 19.324074 to the next highest whole number, 20. This is the *number of courses*.
   3) **(Length)** ⊡ **(number of courses)** ⊟ **(course dimension)**
      260.875 ⊡   20   ⊟ 13.04375
      This is the *course dimension* in decimal inches.
   4) Write down 13.
   5) **(Course dimension)** ⊟ **(whole-number part)** ⊟ ⊠ 16 ⊟ **(no. of 16ths)**
      13.04375   ⊟   13   ⊟ ⊠ 16 ⊟   .7
   6) Round off 0.7 to the nearest whole number: 0.7 rounds off to 1. Add ¹⁄₁₆ to 13, (from step 4). The course dimension is 13 ¹⁄₁₆″.

Example 4   Board-and-battens
You want to side a wall using 1″ × 12″ boards and 1″ × 4″ battens. The siding must have a batten at each end and battens on each side of the door for trim. The wall length is 12′9½″ with a 3′ door located as shown in Figure 2-12. What are the course dimensions of each section so that the courses appear equal? (Figure each section separately.)
   We begin with the section to the left of the door.

A) First, we must find the *length*. Subtract the width of a batten, 3½″, from 5′5″. The length of the section is 5′1½″. Convert the length to decimal inches:
   (Ft.) ⊠ 12 ⊞ **(in.)** ⊞ **(numerator)** ⊡ **(denominator)** ⊟ **(decimal in.)**
   5 ⊠ 12 ⊞ 1 ⊞   1   ⊡   2   ⊟ 61.5
   This is the *length* in decimal inches.

B) The *largest acceptable course* is the width of the board, 11½″, plus the width of the batten, 3½″, minus the overlap for nailing on both sides,

Equally Spaced Courses 23

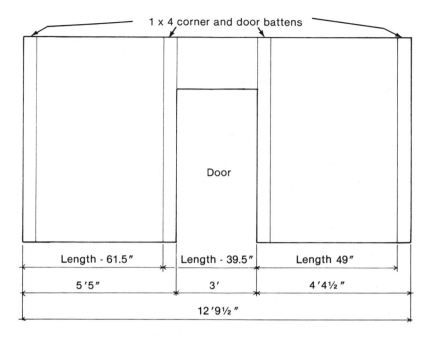

Board-and-batten siding
Figure 2-12

$1\frac{1}{2}''$:
$11.5 \boxplus 3.5 \boxminus 1.5 \boxminus 13.5$
This is the *largest acceptable course* in decimal inches.

C) Now that we have the *length* and the *largest acceptable course,* we are ready to calculate the exact course width.
  1) **(Length)** $\boxdot$ **(largest acceptable course)** $\boxminus$ **(display)**
     61.5 $\boxdot$ 13.5 $\boxminus$ 4.5555556
  2) Round this number up to the next greatest whole number, 5. This is the *number of courses.*
  3) **(Length)** $\boxdot$ **(number of courses)** $\boxminus$ **(course dimension)**
     61.5 $\boxdot$ 5 $\boxminus$ 12.3
     This is the *course dimension* in decimal inches.
  4) Write down 12.
  5) **(Course dimension)** $\boxminus$ **(whole-number part)** $\boxminus$ $\boxtimes$ **16** $\boxminus$ **(no. of 16ths)**
         12.3 $\boxminus$ 12 $\boxminus$ $\boxtimes$ 16 $\boxminus$ 4.8
  6) Round off 4.8 to 5. Add $\frac{5}{16}''$ to 12 (from step 4). The course dimension for the section to the left of the door is $12\frac{5}{16}''$.

24 *Carpentry Layout*

Next, figure the course dimension for the section above the door.

A) First find the *length*. Add the width of a batten, 3½ ", to the width of the door, 3'. *The area to be covered is 3'3½ "*, which we convert to decimal inches.
  (Ft.) ✕ **12** + **(in.)** = **(decimal in.)**
  3 ✕ 12 + 3.5 = 39.5
  This is the *length* in decimal inches.

B) The *largest acceptable course* is the same as before, 13.5".

C) We are now ready to calculate the course dimension.
  1) **(Length)** ÷ **(largest acceptable course)** = **(display)**
     39.5 ÷ 13.5 = 2.9259259
  2) Round this number up to the next nearest greatest whole number, 3. This is the *number of courses*.
  3) **(Length)** ÷ **(number of courses)** = **(course dimension)**
     39.5 ÷ 3 = 13.166667
     This is the *course dimension* in decimal inches.
  4) Write down 13.
  5) **(Course dimension)** − **(whole-number part)** = ✕ **16** = **(no. of 16ths)**
     13.166667 − 13 = ✕ 16 = 2.6666667
  6) Round off 2.6666667 to the nearest whole number, 3. Add ³⁄₁₆" to 13 (from step 4). The course dimension for the area over the door is 13³⁄₁₆".

Finally, we find the course dimension for the area to the right of the door. This area starts with a batten and ends with a batten, so the *length* is the total length of the area to be covered minus the actual width of one batten (see Figure 2-12). We convert the numbers to decimal inches, as usual.

A) First, the length of the area to be covered:
  (Ft.) ✕ **12** + **(in.)** = **(decimal in.)**
  4 ✕ 12 + 4.5 = 52.5
  This is the length of the area to be covered in decimal inches.
  Now subtract the width of the batten, 3½ ", or 3.5":
     52.5 − 3.5 = 49
  This is the *length* in inches.

B) The *largest acceptable course* is still 13.5.

Equally Spaced Courses **25**

C) Next, we calculate the course dimension.
  1) **(Length)** ÷ **(largest acceptable course)** = **(display)**
      49   ÷   13.5   = 3.6296296
  2) Round this number up to the next greatest whole number, 4. This is the *number of courses*.
  3) **(Length)** ÷ **(number of courses)** = **(course dimension)**
      49   ÷   4   = 12.25
     This is the *course dimension* in decimal inches.
 4-6) By now you should recognize that 12.25″ is 12¼″, so the course dimension for the area to the right of the door is 12¼″. You are now ready to layout the boards and battens.

   If you are wondering whether anyone will be able to see the difference between course widths of 12⅚₁₆″, 13³⁄₁₆″, and 12¼″, the answer is no. The courses will look even and you will avoid the cut-up look that extra trim around doors and windows creates.

Example 5   Decking
Suppose you are working for a discriminating owner who requires that you start and end the planking on a sundeck with a full-width board. The deck is 15′5¼″, and the owner will not allow any gap over ³⁄₁₆″ between the 2″ × 6″ planks. What is the course dimension (Figure 2-13)?

A) First, we find the *length,* which is the *length* of the joists to be covered. This is 15′5¼″. Next, convert this to decimal inches:
   (Ft.) × 12 + **(in.)** + **(numerator)** ÷ **(denominator)** = **(decimal in.)**
    15   × 12 +   5   +   1   ÷   4   =   185.25
This is the *length* in decimal inches.

B) Next, find the *largest acceptable course* in decimal inches. This is the width of the plank, 5½″, plus the largest acceptable space between planks, ³⁄₁₆″. Convert these numbers to decimal inches and add them together to find the *largest acceptable course* in decimal inches.
   There are no feet given, so we do not enter anything until we enter inches.
   (Ft.) × 12 + **(in.)** + **(numerator)** ÷ **(denominator)** = **(decimal in.)**
                 5   +   1   ÷   2   =   5.5
Here we start with the fraction, since there is no whole-number part.
                         3   ÷   16   =   .1875
   Now add the two together:
                        5.5   +   .1875   =   5.6875

26 *Carpentry Layout*

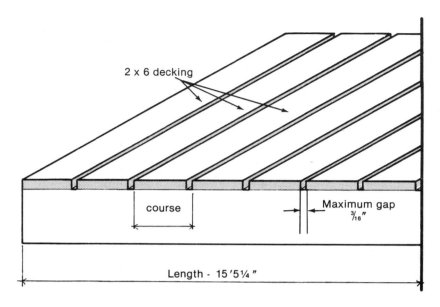

Decking layout
Figure 2-13

This is the *largest acceptable course* in decimal inches.

C) We have the *length* and the *largest acceptable course* in decimal inches. We are now ready to calculate the exact course dimension.
   1) **(Length)** ÷ **(largest acceptable course)** = **(display)**
      185.25 ÷   5.6875   = 32.571429
   2) Round this number to the next greatest whole number, 33. This is the *number of courses*.
   3) **(Length)** ÷ **(number of courses)** = **(course dimension)**
      185.25 ÷   33   = 5.6136364
      This is the *course dimension* in decimal inches.
   4) Write down 5.
   5) **(Course dimension)** − **(whole-number part)** = × 16 = **(no. of 16^{ths})**
      5.6136364   −   5   = × 16 =   9.8181818
   6) Round off 9.8181818 to the nearest whole number: 10. Add $^{10}/_{16}$", or $^{5}/_{8}$", to 5 (from step 4). The course dimension is 5$^{5}/_{8}$".

Example 6 Decking
Again you are working for a demanding owner. But this time, when you measure the deck structure you find one end is exactly 16′ and

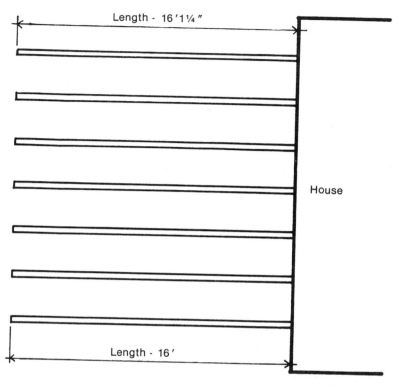

Decking with uneven ends
Figure 2-14

the other end is 16′1¼″. Naturally, the deck must not look out of square when it is done. If you are using 2″ × 6″ planking with a ³⁄₁₆″ maximum gap between planks, what will be your course dimension? (The course dimension will be different at one end than at the other.) (See Figure 2-14.)

A) To solve this problem, we start with the longer end. First, we convert the *length* of the longer side of the deck, 16′1¼″, to decimal inches:
(**Ft.**) ☒ 12 ⊞ (**in.**) ⊞ (**numerator**) ⊡ (**denominator**) ⊟ (**decimal in.**)
16   ☒ 12 ⊞   1   ⊞   1   ⊡   4   ⊟   193.25
This is the *length* of the longer end in decimal inches.

B) Next, we find the *largest acceptable course* in decimal inches. This is the

width of the plank, 5½", plus the space between planks, ³⁄₁₆".
(Numerator) ÷ **(denominator)** + **(in.)** = **(decimal in.)**
   3   ÷   16   +   5.5   =   5.6875
This is the *largest acceptable course* in decimal inches.

C) Now we have the *length* and the *largest acceptable course* in decimal inches. So, we are ready to calculate the course dimension for the longer end.
   1) (Length) ÷ **(largest acceptable course)** = **(display)**
      193.25   ÷   5.6875   =   33.978022
   2) Round 33.978022 up to the next greatest whole number, 34. This is the *number of courses*.
   3) (Length) ÷ **(number of courses)** = **(course dimension)**
      193.25   ÷   34   =   5.6838235
      This is the *course dimension* in decimal inches.
   4) Write down 5.
   5) (Course dimension) − **(whole-number part)** = × 16 = **(no. of 16ths)**
      5.6838235   −   5   = × 16 =   10.941176
   6) Round off 10.941176 to 11. Add ¹¹⁄₁₆" to 5 (from step 4). The course dimension for the long side of the deck is 5¹¹⁄₁₆".

Now, we figure the course dimension for the short side of the deck.

A) First, convert the *length* to decimal inches:
   (Ft.) × 12 = **(decimal in.)**
   16   × 12 =   192
   This is the *length* in decimal inches.

B) We already know the *number of courses* will be 34.

C) We are ready to calculate the exact course dimension, starting at step 3.
   3) (Length) ÷ **(number of courses)** = **(course dimension)**
      192   ÷   34   =   5.6470588
      This is the *course dimension* in decimal inches.
   4) Write down 5.
   5) (Course dimension) − **(whole-number part)** = × 16 = **(no. of 16ths)**
      5.6470588   −   5   = × 16 =   10.352941
   6) Round off 10.352941 to 10. Add ¹⁰⁄₁₆", or ⅝", to 5 (from step 4). The course dimension for the shorter end is 5⅝".

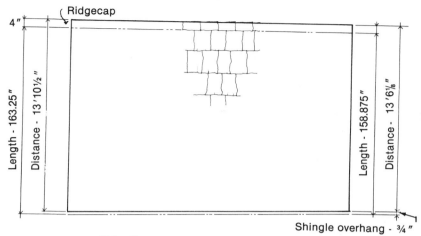

Shingling layout for an uneven roof
Figure 2-15

So, on this deck you will have to gap the planks differently, ¹⁄₁₆″ more at the long end than at the short end. Snap lines on layout with a chalk box every third or fourth course to keep the planks looking straight.

Example 7 Shingles
You are shingling a roof using wood shingles. The ridgecap is 4″ wide, and the shingles must extend ¾″ below the roof sheathing. When you measure the roof, you find that one end is 13′6⅛″ and the other end is 13′10½″. If code won't permit more than 5″ of shingle to the weather, what will be the course dimension (see Figure 2-15)?

First we find the course dimension for the longer end of the roof.

A) We start by finding the *length*. This is the distance from the bottom of the ridgecap course to the bottom of the eaves plus the desired overhang of the shingles at the eaves. Convert 13′10½″, the distance from the top of the ridgecap to the bottom of the roof, to decimal inches:
13 ⊠ 12 ⊞ 10 ⊞ 1 ⊟ 2 ⊟ 166.5
This is the distance in decimal inches.

B) Next, subtract the area covered by the ridgecap, 4″:
166.5 ⊟ 4 ⊟ 162.5
Then, convert the shingle overhang, ¾″, to decimal inches and add it to

the length:
3 ⊕ 4 ⊞ 162.5 ⊟ 163.25
(Note that both of these procedures can be done in one step.) This is the *length* of the long end of the roof.

C) Now calculate the course dimension for the long end.
   1) **(Length)** ⊕ **(largest acceptable course)** ⊟ **(display)**
      163.25 ⊕   5   ⊟ 32.65
   2) Round 32.65 to the next greatest whole number, 33. This is the *number of courses*.
   3) **(Length)** ⊕ **(number of courses)** ⊟ **(course dimension)**
      163.25 ⊕   33   ⊟ 4.9469697
      This is the *course dimension* in decimal inches.
   4) Write down 4.
   5) **(Course dimension)** ⊟ **(whole-number part)** ⊟ ⊠ 16 ⊟ **(no. of 16ths)**
      4.9469697   ⊟   4   ⊟ ⊠ 16 ⊟ 15.151515
   6) Round off 15.151515 to 15. Add $^{15}/_{16}$″ to 4 (from step 4). The course dimension at the long end of the roof is $4^{15}/_{16}$″.

Figuring the course dimension at the short end of the roof is simpler because we know there will be 33 courses. We need only the *length*.

A) First, convert the distance from the top of the ridgecap to the bottom of the roof on the short side, $13′6^{1}/_{8}$″, to decimal inches:
   13 ⊠ 12 ⊞ 6 ⊞ 1 ⊕ 8 ⊟ 162.125
   This is the distance from the top of the ridgecap to the bottom of the roof in decimal inches.

B) Next subtract the area covered by the ridgecap, 4″:
   162.125 ⊟ 4 ⊟ 158.125
   Now add the shingle overhang, ¾″. Convert ¾″ to decimal inches.
   158.125 ⊞ 3 ⊕ 4 ⊟ 158.875
   This is the *length* of the short end of the roof in decimal inches.

C) We know the number of courses will be 33. Therefore we can start at step 3.
   3) **(Length)** ⊕ **(number of courses)** ⊟ **(course dimension)**
      158.875 ⊕   33   ⊟ 4.8143939
      This is the *course dimension* in decimal inches.
   4) Write down 4.
   5) **(Course dimension)** ⊟ **(whole-number part)** ⊟ ⊠ 16 ⊟ **(no. of 16ths)**
      4.8143939   ⊟   4   ⊟ ⊠ 16 ⊟ 13.030303

Equally Spaced Courses   31

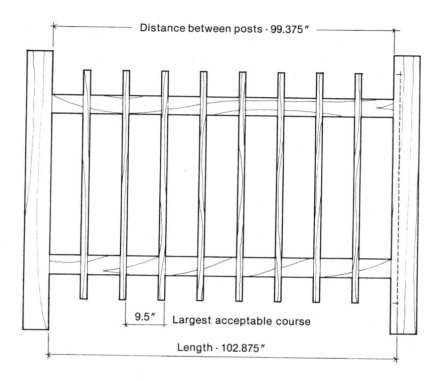

Picket fence layout
Figure 2-16

6) Round off 13.030303 to 13. Add $^{15}/_{16}$" to 4 (from step 4). The course dimension at the short end of the roof is $4^{13}/_{16}$".

The courses at the long side of the roof will be laid out at $4^{15}/_{16}$", while the courses at the short end will be $4^{13}/_{16}$". Snap chalk lines when necessary. Also recalculate the course dimension at about 8 courses from the top so that you can adjust for any error.

Example 8   Pickets
You are building a picket fence. The posts are set at approximately 8' on center. No space greater than 6" between the pickets is allowable. The pickets are 1" × 4", and there will be a space next to the posts (see Figure 2-16). If the distance between the first set of posts (from the inside of one post to the inside of the next post) is 8'3⅜", what will be the course layout?

A) First we find the *length*. In this case, where there is a space next to each post, the *length* is the distance from the inside of one post to the inside of the next post plus the width of one picket. We start by converting the distance between posts, 8′3⅜″, to decimal inches:
8 × 12 + 3 + 3 ÷ 8 = 99.375
This is the distance between posts in decimal inches.
Now, convert 3½″, the actual width of a 1″ × 4″ picket to a decimal, 3.5″. Add that value to the distance between the posts, 99.375″:
99.375 + 3.5 = 102.875
This is the *length* in decimal inches.

B) Now we need the *largest acceptable course*. That is the actual width of a picket, 3½″, plus the largest space allowable, 6″:
3.5 + 6 = 9.5
This is the *largest acceptable course* in decimal inches.

C) Now calculate the course dimension.
   1) **(Length)** ÷ **(largest acceptable course)** = **(display)**
      102.875 ÷   9.5   = 10.828947
   2) Round 10.828947 up to the next greatest whole number, 11. This is the *number of courses*.
   3) **(Length)** ÷ **(number of courses)** = **(course dimension)**
      102.875 ÷   11   = 9.3522727
      This is the *course dimension* in decimal inches.
   4) Write down 9.
   5) **(Course dimension)** − **(whole-number part)** = × 16 = **(no. of 16ths)**
      9.3522727   −   9   = × 16 = 5.6363636
   6) Round off 5.6363636 to 6. Add 6/16″, or ⅜″, to 9 (from step 4). The course dimension is 9⅜″.

Example 9 Handrails
You are building a handrail. The supporting posts measure 9′3½″ from inside to inside. The uprights are 2″ × 2″. Code requires a space less than 9″ between the uprights. An upright is specified next to the supporting post. What is the course dimension (see Figure 2-17)?

   A) As always, we compute the *length* first. In this case, since there are uprights immediately adjacent to the posts, the *length* is the distance from the inside of one post to the inside of the next post minus the width of one upright. To find this value, convert the distance between posts, 9′3½″, to decimal inches:

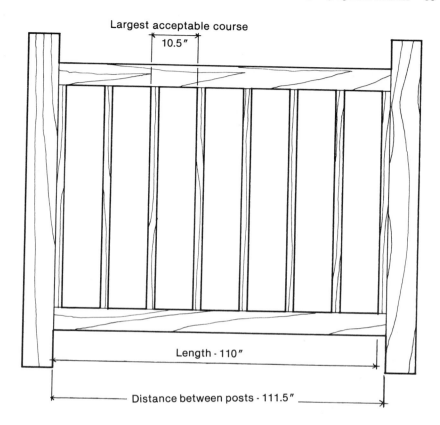

Handrail layout
Figure 2-17

9 ⊠ 12 ⊞ 3.5 ⊟ 111.5
This is the distance in decimal inches.
   Next, subtract the actual width of an upright, 1½", from the distance between the posts. This will give us a *length* to work with.
111.5 ⊟ 1.5 ⊟ 110
This is the *length* in decimal inches.

B) The *largest acceptable course* is the actual width of an upright, 1½", plus the space between uprights, 9". We now convert these to decimal inches and add them together:
1.5 ⊞ 9 ⊟ 10.5
This is the *largest acceptable course* in decimal inches.

C) Now, calculate the course dimension.
  1) **(Length)** ÷ **(largest acceptable course)** = **(display)**
     110   ÷   10.5   =   10.47619
  2) Round 10.47619 up to 11. This is the *number of courses* needed.
  3) **(Length)** ÷ **(number of courses)** = **(course dimension)**
     110   ÷   11   =   10
     This is the *course dimension.*
  4) Since there is no decimal to convert to a fraction, 10″ is the course dimension.

Example 10   Paneling

When working with paneling, you usually cannot adjust course widths because of the presence of tongues and grooves or shiplaps, which must remain intact. Therefore, you should try to make the paneling look symmetrical. You want pieces of the same width at both ends. If you start out with a full-width piece at one end of the wall, you may end up with a narrow piece at the other end. This could be hard to work with, as well as unattractive.

You are paneling an interior wall that is 12′7½″ long with 1″ × 6″ tongue-and-groove redwood paneling placed vertically. What is the width of the first (and last) piece (see Figure 2-18)?

A) The *length* is the length of the area to be covered, 12′7½″. Convert this to decimal inches:
   12 × 12 + 7 + 1 ÷ 2 = 151.5
   This is the wall *length* in decimal inches.

B) Next, find the largest acceptable course, which in this case is the *only* acceptable course, since we do not want to adjust the course width by cutting off tongues and grooves or shiplaps. You can do this fairly accurately by measurement: The face of a 1″ × 6″ panel might measure 5″. Or, you can be very accurate and take the variations of the wood with which you are working into account by using the following method: Cut 10 pieces of paneling to length. Then assemble the cutoff waste on the subfloor, making sure the pieces fit together tightly. Now measure the total length of the 10 pieces and divide by 10. This will give an average width per panel, which is more accurate than the measurement of a face of just one piece. In this case, the total length of all 10 pieces is 51 $^{13}/_{16}$″. Before we can divide by 10, we must convert to decimal inches.
   13 ÷ 16 + 51 = 51.8125
   This is the length of all 10 pieces in decimal inches.

Next, divide by 10 to get the average course width, or *largest acceptable course:*

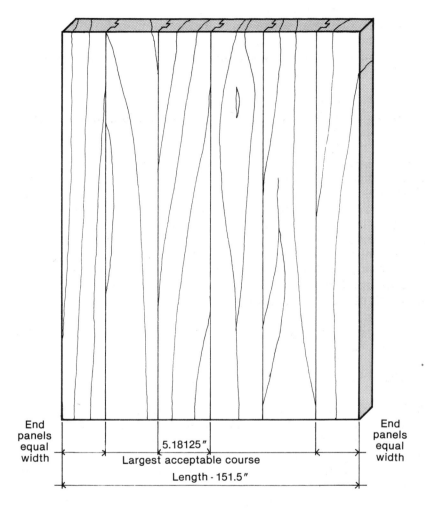

Paneling layout
Figure 2-18

51.8125 ÷ 10 = 5.18125
This is the *largest acceptable course* in decimal inches.

C) Now, we need to know how many courses will cover the wall. To do that, we divide the *length* of the wall by the largest acceptable course (note

that this is the same as step 1):
(Length) ÷ **(largest acceptable course)** = **(number of courses)**
   151.5   ÷   5.18125   =   29.240048
This is the *number of courses*.
This means that we could put on 29 full courses and have a partial course left over to split at each end. However, for both looks and ease of handling, we want the starting and ending panels to be as wide as possible. If we use only 28 full courses, then we have 1.240048 course to split, half at each end. To find the width of each end panel, we divide 1.240048 in half:
   1.240048 ÷ 2 = 0.620024
This is the *portion of a course* with which we will start and end the paneling.
To find the exact width of the first and last panel, multiply the *course width* by the *portion of course,* 0.620024.
(portion of course) × **(average course width)** = **(width)**
   0.620024   ×   5.18125   =   3.2124993
This is the *width of the end panels* in decimal inches.
Last, we convert 3.2124993 to inches and sixteenths of an inch.
Write down 3. This is part of the answer.
3.2124993 − 3 = × 16 = 3.3999896
Round off 3.3999896 to 3, and you have the number of sixteenths, 3. Add this to 3 (from above). The width of an end panel is 3³⁄₁₆".

Now that you have worked through these examples, you can see what sorts of problems can be solved by using the given steps to find equal course widths. You will find other applications during the course of your work. Following are some typical problems you might encounter for you to work out on your own. If necessary, refer back to the appropriate examples for help in finding the answers. The answers are at the end of the book.

Problems

1) You are building a picket fence using $1'' \times 3''$ as pickets. No space larger than $4''$ is allowable, and there will be a space next to the posts. The distance between posts is $6'2\frac{1}{2}''$. Find the course layout.

2) You are paneling a wall with $1'' \times 4''$ tongue-and-groove pine placed vertically. The wall is $18'1\frac{1}{4}''$ long. When you measure 10 pieces of the paneling, you find that they are $32\frac{1}{4}''$ long. What should be the width of the first and last boards?

3) You are working on an exposed rafter ceiling. The rafters are $6'' \times 12''$ and may not exceed $32''$ on center. The length of the house is $56'8''$. What is the course dimension if you wish the rafters to be evenly spaced?

4) You are siding a wall using $1'' \times 10''$ boards and $1'' \times 3''$ battens. The battens must overlap each board by at least $\frac{3}{4}''$ for nailing. The siding will start with a batten and end with a batten. If the wall is $38'$ long, what is the course dimension?

Worksheet

EQUALLY SPACED COURSES

A) Find the length in decimal inches.
   (Ft.) ⊠ 12 ⊞ (in.) ⊞ (numerator) ⊟ (denominator) ⊟ (decimal in.)

B) Find the largest acceptable course in decimal inches.

C) Find the course dimension.
   1) (Length) ⊟ (largest acceptable course) ⊟ (display)
      _____ ⊟ _____ ⊟ _____
   2) Round this number up to the next greatest whole number. This is the number of courses.
   3) (Length) ⊟ (number of courses) ⊟ (course dimension)
      _____ ⊟ _____ ⊟ _____
   4) Write down the whole-number part of the course dimension from step 3.
   5) (Course dimension) ⊟ (whole-number part) ⊟ ⊠ 16 ⊟ (no. of 16ths)
      _____ ⊟ _____ ⊟ ⊠ 16 ⊟ _____
   6) Round off the number from step 5 to the nearest whole number; this gives you the number of sixteenths. Add the fraction to the whole number from step 4. This gives the course dimension.

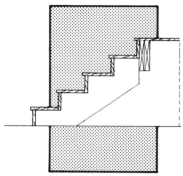

3
Stair Layout

This chapter gives you a reliable system for laying out stairs. You will learn layout for single flights of stairs, stairs with landings, and stairs contoured to the landscape. You can use this system whether the architect has specified certain requirements or whether you are limited only by code. You will be able to calculate all the dimensions needed to cut the stairs: exact rise, exact run, first rise, tread depth, and number of treads (Figure 3-1).

The ideas in this chapter are extensions of those in Chapter 2. You should be able to convert fractions to decimals and back again using your calculator (as described in Chapter 2).

Calculating Stair Layout

To calculate stair layout, you need to know:
- The *height* of the stairs in decimal inches.
- The *length* available for the stairs in decimal inches.
- The *desired rise* in decimal inches.
- The *desired run* in decimal inches, if applicable.
- The *nosing* required in decimal inches.

Stair Layout **39**

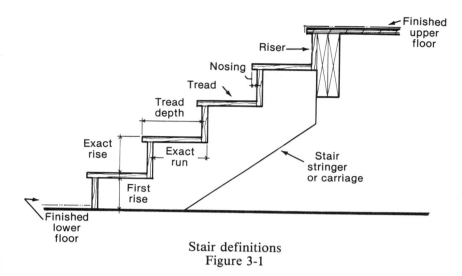

Stair definitions
Figure 3-1

- The *thickness of the subtread* in decimal inches.
- The *thickness of the finish materials* to be applied to the subtreads, the upper subfloor, and the lower subfloor.

TERMS

Now, let's define these terms.

Height.   The **height** of the stairs is the distance from one floor to the next (Figure 3-2). If you do not plan to apply a finish treatment (rugs, tile, hardwood, and so on)—for example on exterior stairs—or if you plan to use the same finish treatment on the upper and lower floors, you need only the measurement from floor to floor. However, if you plan to use a different finish treatment on the upper and lower floors (tile downstairs and carpet upstairs, for example), you must take that into account. This is pretty simple to do: Just take the distance between subfloors, add the thickness of the finish flooring that will be applied to the upper floor, and subtract the thickness of the finish flooring that will be applied to the lower floor. You can do this on a calculator, as shown later.

Desired Rise.   The **desired rise** is the height that you or the architect would like to have from one tread to the next.

Length.   The **length** is the maximum length available for the stairs (Figure 3-3). Often you will not need to use the full space available for the stairs, but occasionally you must use the full space; in that case the length is fixed.

40 *Carpentry Layout*

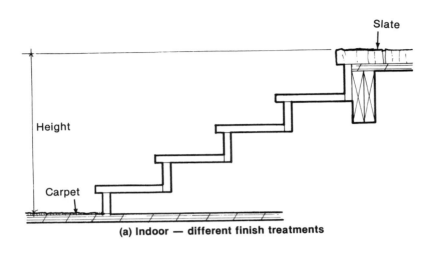

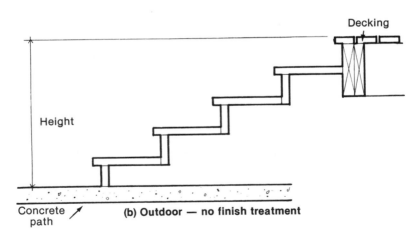

Stair heights
Figure 3-2

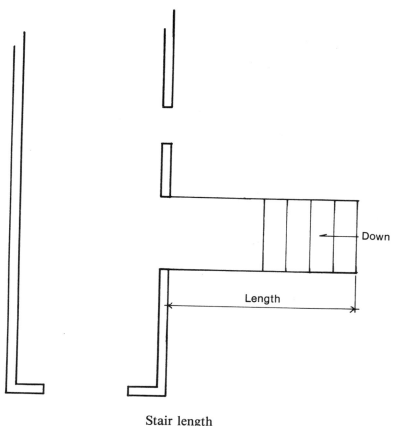

Stair length
Figure 3-3

Desired Run. The **desired run** is the depth of the tread (less the nosing) that you or the architect would like to have from one riser to the next. If you must work with a fixed length, then there is no desired run. You must use the run that works in that space.

Nosing. The **nosing** is the amount of overhang on the front of a tread.

You will find that the first rise is usually different from the exact rise. This is because you must allow for the difference in the thickness of the floor and tread finishes.

The *desired rise* and the *desired run* are ideal dimensions only. Because of the many factors involved in stair layout, you'll have to adjust these figures to find *exact rise* and *exact run*—dimensions that will actually work for a given set of stairs.

So, first you must find out all these measurements. Then you are ready to calculate the stair layout. Let's first go over the entire process. Begin by finding the *exact rise*. First divide the *height* by the *desired rise* and round off the answer to the nearest whole number. This gives the number of risers you will use. Now divide the height by the number of risers to get the *exact rise* in decimal inches. Convert to fractional inches for your answer.

Now calculate the *number of treads* and the *exact run*. Subtract 1 from the *number of risers* for the *number of treads* you will use. Then divide the *length* by the *number of treads*. This will give you the largest possible run you could use on these stairs. The *exact run* will be either the largest possible run or the desired run, whichever is smaller. If there is no desired run, then the largest possible run is the same as the exact run.

To find the tread depth, add the *exact run* to the *nosing* dimension.

Finally, you must find the first rise. Begin with the exact rise. Subtract the subtread thickness, subtract the finish tread thickness, and add the thickness of the finish flooring for the lower floor. If you plan to give both the stairs and the lower floor the same finish—hardwood, for example—then just subtract the subtread thickness from the exact rise.

Beginning Work

Here are the instructions for using a calculator.

A) Find the height of the stairs.
  1) **(Distance between subfloors)** ⊞ **(thickness of finish on upper floor)** ⊟ **(thickness of finish on lower floor)** ⊟ **(height)**

B) Then, find the *number of risers* and the *exact rise*.
  1) **(Height)** ⊡ **(desired rise)** ⊟ **(display)**
  2) Round the number on display to the nearest whole number. This is the number of risers you will use.
  3) **(Height)** ⊡ **(number of risers)** ⊟ **(exact rise in decimal in.)**
  4) Convert decimal inches to fractional inches to get the exact rise. See Chapter 2 for this procedure.

C) Third, calculate the *number of treads* and the *exact run*.

1) **(Number of risers from Step B2)** ⊟ 1 ⊟ **(number of treads)**
2) **(Length)** ⊟ **(number of treads)** ⊟ **(largest possible run in decimal inches)**
3) The exact run is the smaller of the desired run or the largest possible run.
4) Convert the answer to fractional inches.

D) Next, find the *tread depth*.
  1) **(Exact run)** ⊞ **(nosing)** ⊟ **(tread depth)**
  2) Convert the answer to fractional inches.

E) Finally, find the *first rise*.
  1) **(Exact rise)** ⊟ **(subtread thickness)** ⊟ **(thickness of finish tread)** ⊞ **(thickness of lower-floor finish)** ⊟ **(first rise in decimal inches)**
  2) Convert the answer to fractional inches.

You now have all the information necessary to lay out a set of stairs: exact rise, number of treads, exact run, tread depth, and first rise.

Applications

Let's look at some practical examples using this method.

Example 1

Let's suppose you are going out on your first job. The stairs are very straightforward—as will be at least half the stairs you lay out. You have been given all the dimensions necessary to lay out a set of stairs from the first floor to the second floor as follows (see Figure 3-4):

- The distance between subfloors is 9′3½″.
- The length available for the stairs is 18′.
- The architect desires a 7½″ rise and a 10½″ run with a ¾″ nosing.
- Subtreads are to be cut from 1⅛″ plywood and are to be finished with ⅞″-thick carpet.
- The first and second floors will be covered with ⅞″-thick carpet.

A) First, you must figure the *height* of the stairs. The distance between subfloors is 111.5″.
  1) **(Distance between subfloors)** ⊞ **(thickness of the carpet on the upper floor)** ⊟ **(thickness of the carpet on the lower floor)** ⊟ **(height of the stairs)**
     111.5 ⊟ .875 ⊞ .875 ⊟ 111.5

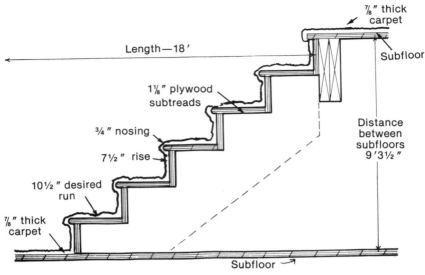

Straightforward stairs
Figure 3-4

(*Note:* You can see that since both floors get the same finish, the distance between subfloors is the same as the height of the stairs.)

B) Find the number of risers and the exact rise:
1) **(Height)** ÷ **(desired rise)** = **(display)**
   111.5 ÷ 7.5 = 14.866667
2) Round the number on display to the nearest whole number. This is the number of risers you will use: 14.866667 rounds to 15.
3) **(Height)** ÷ **(number of risers)** = **(exact rise)**
   111.5 ÷ 15 = 7.4333333
4) Convert to fractional inches: 7.4333333″ = 7 7/16″.

C) Now calculate the number of treads and the exact run.
1) **(Number of risers)** − 1 = **(number of treads)**
   15 − 1 = 14
2) **(Length)** ÷ **(number of treads)** = **(largest-possible run)**
   216 ÷ 14 = 15.428571
3) Select the largest-possible run or the desired run, whichever is smaller: 10.5″ is smaller than 15.428571″.
4) Convert 10.5″ to fractional inches: 10.5″ = 10 1/2″. This is the exact run.

D) Next, find the tread depth.
  1) **(Exact run)** ⊞ **(nosing)** ⊟ **(tread depth)**
     10.5   ⊞   .75   ⊟   11.25
  2) Convert the tread depth to fractional inches: 11.25″ = 11¼″.

E) Last, figure the first rise.
  1) **(Exact rise)** ⊟ **(subtread thickness)** ⊟ **(first rise)**
     7.4333333   ⊟   1.125   ⊟   6.3083333
     (Note: Since both the finish tread and the first-floor finish are the same, you just subtract the subtread thickness, 1.125″, from the exact rise to get the first rise.)
  2) Convert the first rise to fractional inches: 6.3083333″ = 6$\frac{5}{16}$″.

Before you begin work, summarize the dimensions you will need to lay out this set of stairs:
- Exact rise = 7$\frac{7}{16}$″
- Number of treads = 14
- Exact run = 10½″
- Tread depth = 11¼″
- First rise = 6$\frac{5}{16}$″

Example 2
Now, let's consider a job with a little more challenge. You have talked to the owner and to the architect. Now you must find all the dimensions necessary to lay out a set of stairs from the first floor to the second floor from your information (Figure 3-5):
- The height from the first to the second subfloor is 11′ 3½″. The finished floor on the first floor will be slate 3″ thick. The second-floor finish will be hardwood flooring ¾″ thick.
- The architect desires a 6″ rise and a 12″ run with no nosing.
- The subtreads are to be from 2″ nominal material and will get ¾″ hardwood finish.
- The length available for the stairs is 20′4″.

A) First, figure the height from finished floor to finished floor.
  1) **(Distance between subfloors)** ⊞ **(thickness of finish on upper floor)** ⊟ **(thickness of finish on lower floor)** ⊟ **(height)**
     135.5 ⊞ .75 ⊟ 3 ⊟ 133.25

B) Next, figure the exact rise:
  1) **(Height)** ⊡ **(desired rise)** ⊟ **(display)**
     133.25   ⊡   6   ⊟   22.208333

46 *Carpentry Layout*

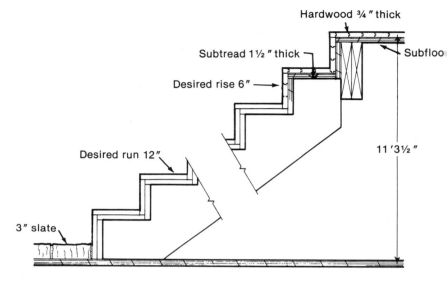

More challenging stairs
Figure 3-5

2) Round the number on display to the nearest whole number.
This is the *number of risers* you will use: 22.208333 rounds to 22.
3) **(Height)** ÷ **(number of risers)** = **(exact rise)**
    133.25 ÷   22   = 6.0568182
4) Convert the exact rise to fractional inches: 6.0568182″ = 6 1/16″.

C) Find the number of treads and the exact run.
1) **(Number of risers)** − 1 = **(number of treads)**
      22   − 1 = 21
2) **(Length)** ÷ **(number of treads)** = **(largest-possible run)**
      244   ÷   21   = 11.619048
3) Select the largest-possible run, 11.619048″, or the desired run, 12″, whichever is smaller, for exact run. Since 11.619048″ is smaller, it is the exact run.
4) Convert 11.619048 to fractional inches: 11.619048″ = 11 5/8″.

D) Next, find the tread depth.
1) **(Exact run)** + **(nosing)** = **(tread depth)**
    11.619048 +   0   = 11.619048
2) Convert the tread depth to fractional inches: 11.619048″ = 11 5/8″.

E) Last, find the first rise.
  1) **(Exact rise)** ⊟ **(subtread thickness)** ⊟ **(thickness of finish tread)** ⊞ **(thickness of lower-floor finish)** ⊟ **(first rise)**
     6.0568182 ⊟ 1.5 ⊟ .75 ⊞ 3 ⊟ 6.8068182
  2) Convert the first rise to fractional inches: 6.8068182″ = 6 13/16″.

Again, summarize your answers before you begin work:
- Exact rise = 6 1/16″
- Number of treads = 21
- Exact run = 11 5/8″
- Tread depth = 11 5/8″
- First rise = 6 13/16″

Example 3

You need to build a set of stairs from a parking pad to a path, contoured to the landscape and set in dirt. You are required to build a frame for each step, which will be filled with asphalt or gravel (Figure 3-6). Find all the dimensions necessary to lay out the stairs.

The first dimensions you must find are the *height* and the *length*.

By measuring, you find the height from the parking pad to the path to be 4′8 1/2″. Measuring on the level, the length is 14′8″ from the end of the parking pad to the beginning of the path. The length cannot be shorter or it will not reach the landing. Because the length is fixed, there is no desired run. You would like a rise of 6″.

A) You found the height to be 4′8 1/2″.

B) Figure the exact rise.
  1) **(Height)** ⊟ **(desired rise)** ⊟ **(display)**
     56.5 ⊟ 6 ⊟ 9.4166667
  2) Round the number on display to the nearest whole number.
     This is the number of risers you will use: 9.4166667 rounds to 9.
  3) **(Height)** ⊟ **(number of risers)** ⊟ **(exact rise)**
     56.5 ⊟ 9 ⊟ 6.2777778
  4) Convert the exact rise to fractional inches: 6.2777778″ = 6 1/4″.

C) Next, figure the number of treads and the exact run.
  1) **(Number of risers)** ⊟ 1 ⊟ **(number of treads)**
     9 ⊟ 1 ⊟ 8
  2) **(Length)** ⊟ **(number of treads)** ⊟ **(largest-possible run)**
     176 ⊟ 8 ⊟ 22

48 *Carpentry Layout*

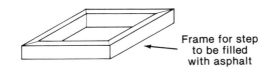

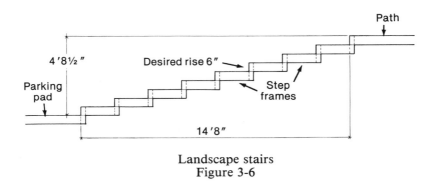

Landscape stairs
Figure 3-6

3) Because the length is fixed, the largest-possible run is the exact run.

So, the rise is 6¼ ", the run is 22 ", and the number of treads (in this case, boxes) is 8. Because no finish material is to be applied on top of the risers, the tread depth is 22 " and the first rise is the same as the others.

Again, before you begin to build the boxes, summarize your answers:
- Exact rise = 6¼ "
- Number of treads (boxes) = 8
- Exact run = 22"
- Tread depth = 22"
- First rise = 6¼ "

Example 4
You need to lay out a set of exterior entry stairs from a concrete walk to a front porch. You have measured the area and talked to the architect. Now you are ready to find all the dimensions necessary from your information (Figure 3-7):

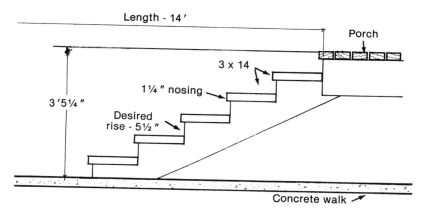

Exterior entry stairs
Figure 3-7

- The height from the concrete walk to the top of the porch is 3'5¼".
- The length available is 14'.
- The treads are to be 3 × 14s (actual measurements: 2½" × 13½").
- The desired rise is 5½".
- 1¼" nosing is required.

A) You already have the height by measurement, so you are ready to begin the calculation. The height from finished walk to finished porch is 41¼".

B) Next, figure the exact rise.
   1) (Height) ÷ **(desired rise)** = **(display)**
       41.25 ÷   5.5   = 7.5
   2) Round to the nearest whole number.
   This is the number of risers: 7.5 rounds to 8.
   3) (Height) ÷ **(number of risers)** = **(exact rise)**
       41.25 ÷   8   = 5.15625
   4) Convert the exact rise to fractional inches: 5.15625" = 5³⁄₁₆".

C) Now, figure the number of treads and the exact run.
   1) (Number of risers) − 1 = **(number of treads)**
       8   − 1 = 7

2) (Length) ÷ (number of treads) = (largest-possible run)
   168   ÷   7   =   24
3) Since you know only the material for the treads, 3 × 14s (2½" × 13½"), you must find the desired run: That is the tread depth less the amount needed for the nosing, or 13½" − 1¼" = 12¼".

   Select the smaller of the largest possible run (24") or the desired run (12¼"): 12¼" is the exact run.
4) The answer is already in fractional form.

D) The treads are to be cut from 3 × 14s, so the tread depth must be 13½".

E) The last dimension you need to find is the first rise.
   1) (Exact rise) − (subtread thickness) − (thickness of finish tread) + (thickness of lower-floor finish) = (first rise)
      5.15625 − 2.5 − 0 + 0 = 2.65625
   2) Convert the first rise to fractional inches: 2.65625" = 2¹¹⁄₁₆".

Summarize your answers:
- Exact rise = 5³⁄₁₆"
- Number of treads = 7
- Exact run = 12¼"
- Tread depth = 13½"
- First rise = 2¹¹⁄₁₆"

Example 5
Now that you are more experienced, you have been hired to put in a set of stairs with a landing (Figure 3-8). Find all the dimensions necessary to lay out the stairs and landing if you have the following information:
- The height from subfloor to subfloor is 9′2".
- The desired rise is 7¾".
- The desired run is 10½".
- The floors and stairs will all be carpeted with the same material.
- The subtread is to be cut from 1⅛" plywood.
- Nosing is to be 1".
- The length for the top flight of stairs is 89" less 32" for the landing (or 57"), and the available length for the lower flight is 10′.

There are two special problems in figuring the layout for stairs with landings. First, you must consider the length to be fixed on the top flight of stairs: If the stairs are too short, they won't reach the landing. If they are too long, they will stick out onto the landing—and that is a possible code violation. Second, you need to determine

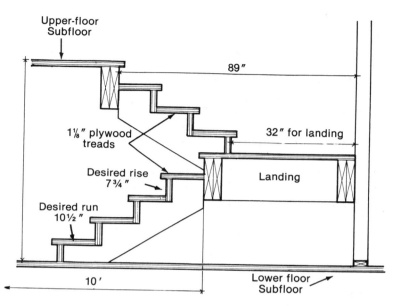

Stairs with a landing
Figure 3-8

where to put the landing, for it must be built before the stairs.

Start by finding the exact rise. Since the exact rise and the exact run will be the same on both flights of stairs, you can figure the exact rise as if you were figuring one flight.

A) The height from subfloor to subfloor is 110″. This is also the height from finished floor to finished floor.

B) Now, calculate the exact rise:
  1) (Height) ÷ (desired rise) = (display)
     110   ÷   7.75   =   14.193548
  2) Round the number to the nearest whole number.
     This is the *number of risers:* 14.193548 rounds to 14.
  3) (Height) ÷ (number of risers) = (exact rise)
     110   ÷   14   =   7.8571429
  4) Convert the exact rise to fractional inches: 7.8571429″ = 7⅞″.

C) Next, you need to find the total number of treads and exact run.

1) **(Number of risers)** ⊟ 1 ⊟ **(number of treads)**
   14   ⊟ 1 ⊟ 13

The landing itself is a tread, so that leaves 12 treads to divide up between the top and bottom flights of stairs.

Now, you must determine the exact run. Since the length is fixed, the exact run will be the same as the largest-possible run. The length for the top flight of stairs is 57", but you don't know how many treads the top flight has. The number of risers is found by dividing the *height* by the *desired rise*. Similarly, the number of treads is found by dividing the *length* by the *desired run*. To find this on your calculator:

- **(Length)** ⊟ **(desired run)** ⊟ **(display)**
   57   ⊟   10.5   ⊟   5.4285714
- Round the number on display to the nearest whole number: 5.4285714 rounds to 5, which is the number of treads.

2) **(Length)** ⊟ **(number of treads)** ⊟ **(exact run)**
   57   ⊟   5   ⊟   11.4

3) Because the length is fixed, you must use 11.4" as the exact run.

4) Convert the run to fractional inches: 11.4" = 11⅜", the exact run for both flights of stairs.

Now you can determine where to put the landing. Since the top flight of stairs has 5 treads, it has 6 risers. You know the exact rise is 7.8571429, so:

- **(Exact rise)** ⊠ **(number of risers)** ⊟ **(height of top flight of stairs)**
   7.8571429 ⊠   6   ⊟   47.142857
- Convert the height to fractional inches: 47.142857" = 47⅛".

So, you must build the finished floor of the landing 47⅛" below the finished second floor. Also, the bottom flight of stairs will have 7 treads (12 treads less 5 treads in the top flight of stairs).

D) Now, find the tread depth.
   1) **(Exact run)** ⊞ **(nosing)** ⊟ **(tread depth)**
      11.4   ⊞   1   ⊟   12.4
   2) Convert the tread depth to fractional inches: 12.4" = 12⅜".

E) Finally, find the first rise. Since the floors, landing, and stairs all receive the same finish treatment, you need only subtract the thickness of the subtread from the exact rise to find the first rise.

   1) **(Exact rise)** ⊟ **(subtread thickness)** ⊟ **(first rise)**
      7.8571429 ⊟   1.125   ⊟   6.7321429
   2) Convert the first rise to fractional inches: 6.7321429" = 6¾"

Problems

1) Find all the dimensions necessary to lay out a set of stairs given the following information:
   - The distance between subfloors is 12′3½″.
   - The length available for the stairs is 21′.
   - The architect desires a 7″ rise and an 11″ run with a ¾″ nosing.
   - Subtreads are to be cut from 1⅛″ plywood.
   - The first floor, second floor, and stairs will be covered with carpet and pad, which are 1¼″ thick in all.

2) Find all the dimensions necessary to lay out a set of stairs given the following information:
   - The distance between subfloors is 10′1½″.
   - The length available for the stairs is 18′8″.
   - The architect desires a 6″ rise and a 12″ run with a 1″ nosing.
   - Subtreads will be 1½″ thick.
   - The first floor will be covered with 3½″-thick slate; the second floor will be covered with 1¼″-thick carpeting; and the stairs will be covered with ½″-thick tile.

3) You need to put in a set of stairs with a landing. Find all dimensions necessary to lay out the stairs and landing given the following information:
   - The height from subfloor to subfloor is 14′8″.
   - The desired rise is 7½″.
   - The desired run is 11″.
   - The floors and stairs will be carpeted with the same material.
   - The subtreads will be 1½″ thick.
   - The nosing will be ¾″.
   - The length for the top flight of stairs is 80½″. The length available for the lower flight is 22″.

Worksheet

STAIRS

A) Find the height of the stairs:
1) **(Distance between subfloors)** ⊞ **(thickness of finish on upper floor)** ⊟ **(thickness of finish on lower floor)** ⊟ **(height of stairs on display)**

   _____ ⊞ _____ ⊟ _____ ⊟ _____

B) Then, find the exact rise.
1) **(Height)** ⊟ **(desired rise)** ⊟ **(display)**

   _____ ⊟ _____ ⊟ _____
2) Round the number on display to the nearest whole number. This is the number of risers you will use.
3) **(Height)** ⊟ **(number of risers)** ⊟ **(exact rise in decimal inches)**

   _____ ⊟ _____ ⊟ _____
4) Convert decimal inches to fractional inches to get the exact rise.

C) Next, calculate the number of treads and the exact run.
1) **(Number of risers)** ⊟ 1 ⊟ **(number of treads)**

   _____ ⊟ 1 ⊟ _____
2) **(Length)** ⊟ **(number of treads)** ⊟ **(largest-possible run)**

   _____ ⊟ _____ ⊟ _____
3) The exact run is the smaller of the desired run or the largest-possible run.
4) Convert the answer to fractional inches.

D) Find the tread depth.
1) **(Exact run)** ⊞ **(nosing)** ⊟ **(tread depth)**

   _____ ⊞ _____ ⊟ _____
2) Convert the answer to fractional inches.

E) Find the first rise.
1) **(Exact rise)** ⊟ **(subtread thickness)** ⊟ **(thickness of finish tread)** ⊞ **(thickness of lower-floor finish)** ⊟ **(first rise)**

   _____ ⊟ _____ ⊟ _____ ⊞ _____ ⊟ _____
2) Convert the answer to fractional inches.

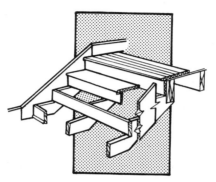

4
Stair Layout in the Field

Suppose you have calculated a set of stairs, as in the first example of Chapter 3. Now you must build the stairs. What do you do next?

Laying Out the Stairs

To lay out and cut a set of stairs, you will need the following:
1. A rafter square with markers (Figure 4-1). This is a square which is 16″ long on the tongue, or short side, and 24″ long on the body, or long side.
2. A saw.
3. Information about the stairs. We'll use the figures calculated in Example 1 of Chapter 3.
    - Exact rise = $7\frac{7}{16}''$
    - Number of treads = 14
    - Exact run = $10\frac{1}{2}''$
    - Tread depth = $11\frac{1}{4}''$
    - First rise = $6\frac{5}{16}''$

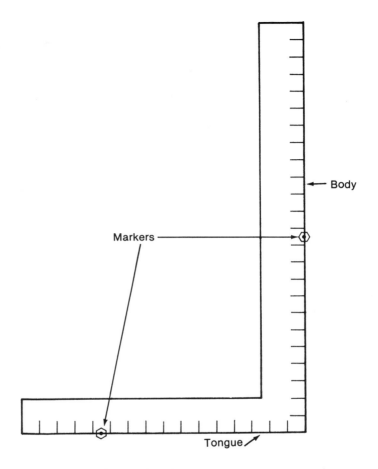

Rafter or framing square
Figure 4-1

Stair Layout in the Field 57

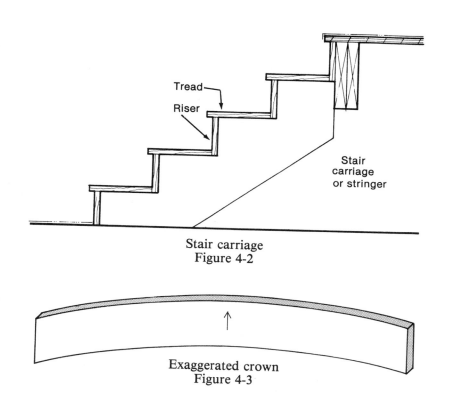

Stair carriage
Figure 4-2

Exaggerated crown
Figure 4-3

Let's say this set of stairs is 32" wide and requires two carriages.

STAIR CARRIAGES

First choose the lumber to use for the stair carriages. The **stair carriages** are the supporting members for the treads and risers (Figure 4-2). "Crown" the carriages by sighting along the edge of the lumber and drawing an arrow pointing toward the edge with a hump in it (Figure 4-3). Don't use less than 2 × 12 structurally graded lumber.
Marking Tread Lines. Place the two carriages on a pair of saw horses with the crown away from you. This will help you visualize the stairs better. Set the markers on your framing square at 7⁷⁄₁₆" on the tongue and 10½" on the body. Place the rafter square as shown in Figure 4-4 and trace the outline of it with your pencil. It is important to visualize the stair carriage in place to keep from making mistakes. Now slide the rafter square down until the marker at 7⁷⁄₁₆" lines up

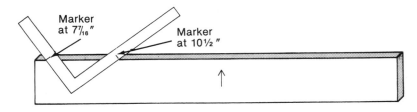

Beginning layout
Figure 4-4

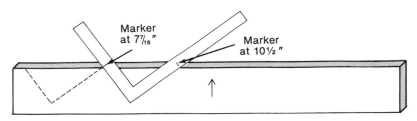

Continuing layout
Figure 4-5

with the line you drew along the body of the square (Figure 4-5). Repeat this process until you count 14 tread lines (these are the lines drawn along the body of the square).

Top and Bottom Cuts. Now you have to draw the top and bottom cuts for the carriages. At the bottom of the carriage, draw a riser line as usual. Then measure the first rise distance, $6\frac{5}{16}"$, from the tread line, and place a mark there (Figure 4-6). Slide the rafter square along the carriage until the body lines up with this mark (Figure 4-7). Draw a line along the body of the square and extend it through the carriage. This is the bottom cut.

For the top cut, extend the riser line that runs above the top tread through the carriage (Figure 4-8).

Cutting the Carriages. To cut the carriages, use either a handsaw or a circular saw; however, if you use a circular saw, finish your cuts with a handsaw to keep from weakening the carriages by cutting too deeply.

Cutting the Treads and Risers. The treads and risers can now be cut. The treads can all be the same—$11\frac{1}{4}" \times 32"$. All the risers are the same—$7\frac{7}{16}" \times 32"$—except for the first riser and the last riser. The

Stair Layout in the Field **59**

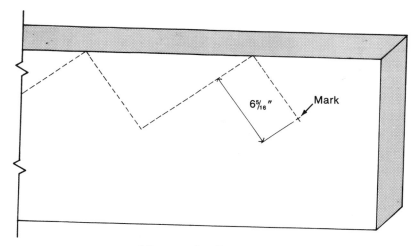

Measure for first riser
Figure 4-6

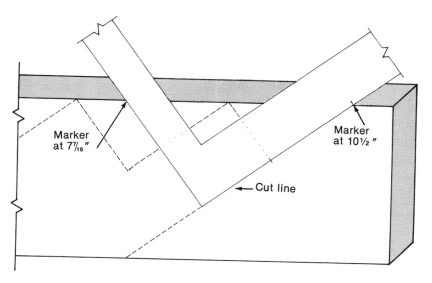

Draw cut line at first riser
Figure 4-7

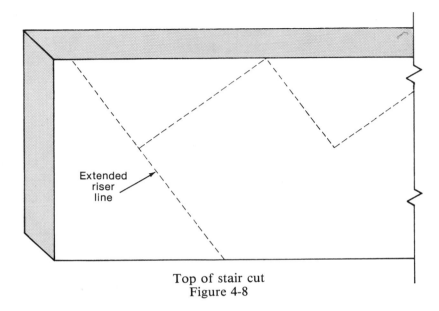

Top of stair cut
Figure 4-8

first riser is 6⁵⁄₁₆" × 32" to allow for the special first rise. The top riser depends upon the thickness of the floor material on the upper floor (Figure 4-9).

Installing the Stairs

When installing the stairs, measure down from the upper floor. Measure distance that allows for the subtread to be placed on the carriage. If the stairs and the upper floor are to receive the same finish materials, measure down from the upper subfloor the distance of the exact rise—7⁷⁄₁₆"—plus the subtread thickness—1⅛" (Figure 4-10). The top of the stair carriage will line up with this mark to give you a level set of stairs with the proper risers and treads.

When installing the treads and risers, put the risers in first. This will allow you to nail through the upper riser into the tread, and you can glue the tread onto the lower riser (Figure 4-11).

Stair Layout in the Field 61

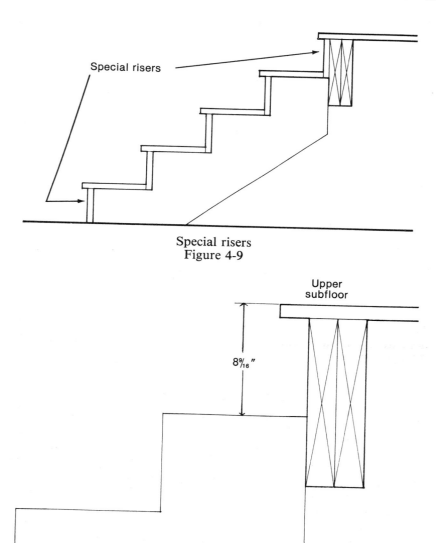

Special risers
Figure 4-9

Measuring to top of carriage for installation
Figure 4-10

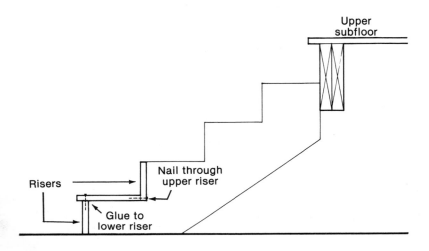

Installing treads and risers
Figure 4-11

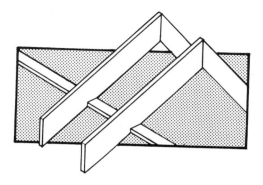

5
Common Rafter Layout

In this chapter you learn how to calculate the layout of a common rafter for a roof of any pitch. There are many books full of tables for figuring rafter lengths, but often these books do not cover the exact situation on which you are working. Interpreting the answers given in the tables can often be confusing and lead to inaccurate layouts. Using the system described in this chapter will give you exact answers for all the layout problems you might encounter with common rafters.

TERMS

To avoid confusion, you'll need to know some roofing terms.

Ridge Pole. The highest horizontal roof member is the **ridge pole**. The rafters are connected to it (Figure 5-1).

Ridge Line. Sometimes there is no ridge pole; the rafters are tied to one another at the top. In this case the highest point of the roof (the peak) is called the **ridge line** (Figure 5-2).

Rafters. **Rafters,** the members that support the sheathing, can be of several types: common, hip, valley, jack and cripple jack. **Common rafters,** with which we are concerned in this chapter, are rafters that run at right angles from the ridge pole or ridge line to the edge of the

64 *Carpentry Layout*

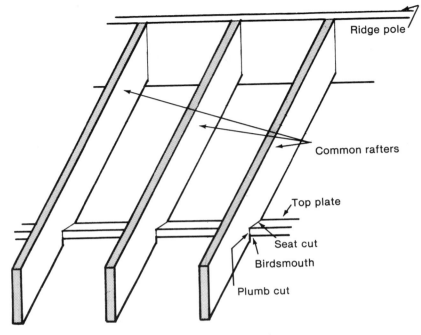

Roof framing definitions
Figure 5-1

building, as in Figure 5-1. The other kinds of rafters will be discussed in later chapters.

Birdsmouth. The **birdsmouth** is the cut out in the rafter which fits over the supporting beam or wall (Figure 5-3). The vertical cut is called the **plumb cut of the birdsmouth.** The level cut is called the **seat cut.**

WHAT YOU'LL NEED TO KNOW

The layout you'll calculate includes the length of the rafter, the length of the overhang, and the placement of the birdsmouths. To calculate these figures, there are several dimensions you must know. These may be found on the plans or taken directly from the building:

- The *unit rise* of the roof.
- The *run of the overhang* in decimal inches.
- The *run of the rafter* in decimal inches.
- The *run between birdsmouth plumb cuts,* if applicable, in decimal inches.
- The *seat cut* required at the birdsmouth in decimal inches.

Common Rafter Layout 65

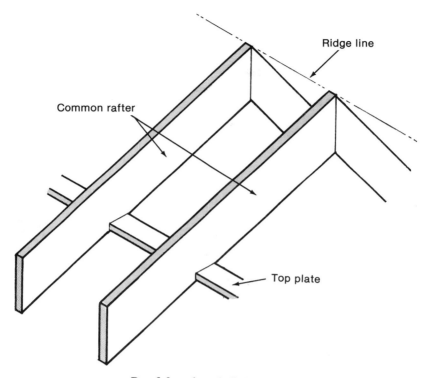

Roof framing definitions
Figure 5-2

Unit Rise. The **unit rise** of the roof is given in the plans as the upper number (or first number) in the pitch. For example, if the pitch is 4/12 or 4:12, the unit rise is 4. This is the distance, in inches, that the roof rises for every 12″ in level, or horizontal, distance. In this fraction, or ratio, 12 is the **unit run**.

Run of the Overhang. The **run of the overhang** is the level distance the rafter hangs over the outside of the building (Figure 5-4). This dimension is found on the roof plan or an elevation.

Run of the Rafter. The **run of the rafter** is the level distance that the rafter covers. There are two situations you'll have to consider here, a shed roof and a gable roof. A shed roof is one sloping surface (Figure 5-5). A gable roof has two sloping surfaces, which join at the ridge. The run of a rafter for a shed roof is the width of the building plus the run of the overhangs. The run of a rafter for a gable roof is the level distance that the rafter runs, including the rafter overhang but not the

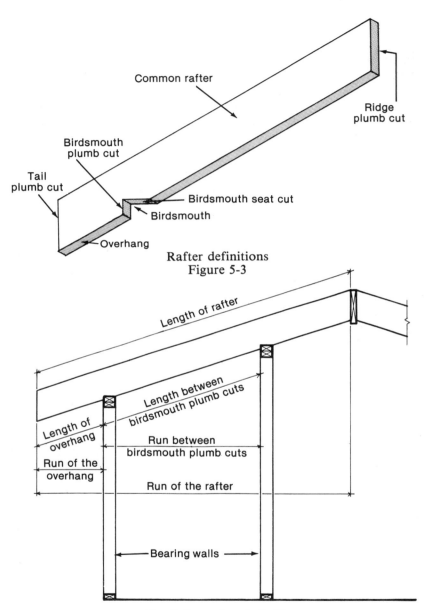

Rafter definitions
Figure 5-3

Roof dimension definitions
Figure 5-4

ridge pole.

The ridge pole is usually centered on the building. Thus to find the run of the rafter for a gable roof, you must divide the width of the building (in decimal inches) in half. This gives you a dimension to the center of the ridge pole. Next you subtract half the width of the ridge pole; this gives us the run of the rafter inside the building. Finally, you add the run of the overhang to arrive at the run of the rafter. On a calculator, follow this sequence:

A) Find the run of a rafter for a gable roof.*
   1) Convert the building width to decimal inches.
      (Ft.) ☒ 12 ⊞ **(in.)** ⊞ **(numerator)** ⊟ **(denominator)** ⊟ **(building width in decimal inches)**
   2) Divide the width of the building in half.
      (Building width) ⊟ 2 ⊟ **(half of width)**
   3) Divide the width of the ridge pole in half and subtract it from half the building width.
      (Half of building width) ⊟ **(ridge pole)** ⊟ 2 ⊟ **(run of rafter inside the building)**
   4) Add the run of the overhang.
      (Run of rafter inside the building) ⊞ **(run of overhang)** ⊟ **(run of rafter)**

Run Between Birdsmouth Cuts. The **run between birdsmouth plumb cuts** is the level distance from the plumb cut of one birdsmouth to the plumb cut of the next (Figure 5-4). This dimension can be measured on the subfloor from the outside of one bearing wall to the outside of the next. If a bearing beam is used instead of a bearing wall, you can measure from the outside of the supporting post to the outside of the next supporting post or bearing wall to find the *run between birdsmouth plumb cuts*. Many rafters will have only one bearing wall; in that case, this dimension can be ignored.

Seat Cut. The **seat cut** required at the birdsmouth can be full bearing or partial bearing (Figure 5-6). Usually the rafter gets **full bearing**; in that case the seat cut of the birdsmouth is the same dimension as the thickness of the wall on which it bears. Occasionally **partial bearing** is

---

* You will observe as you use these calculations that you may often be able to go through the entire sequence of related step without clearing your calculator, using the answer from one as the first entry in the next.

68 *Carpentry Layout*

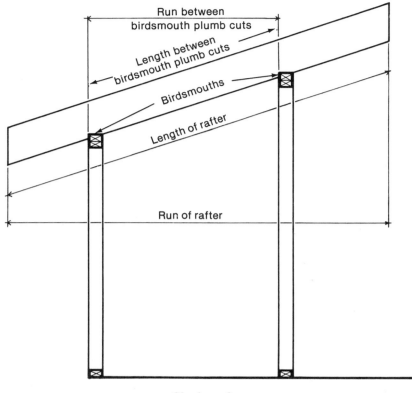

Shed roof
Figure 5-5

required. This usually happens when the wall or beam on which the rafter bears is so thick that giving the rafter full bearing would cut the rafter too deeply to retain its strength. In this case the bearing, which is the same as the **seat cut,** is given in the plans.

Beginning Work

Before starting to calculate, draw a picture of the rafter with the dimensions you know (in decimal inches) filled in and the dimensions you must find left blank (Figure 5-7). This will help you visualize the rafter and keep you from getting confused. When it's drawn in this way, you can give it to someone else to cut without error.

Now you are ready to calculate the remaining dimensions. You

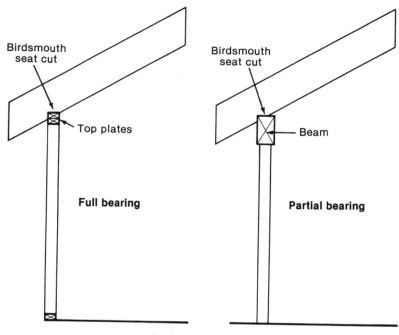

Rafter bearing
Figure 5-6

are going to find a number, the *rake multiplier,* which you can multiply by the run, or level distance, of the different parts of the rafter to give you the length of those parts. To get this number, you square the *unit rise* of the roof. Next, you square the unit run, which is always 12, and add it to the square of the unit rise. Then you take the square root of that sum and divide by 12. This number is the rake multiplier. This number is different for each different roof pitch. By multiplying the *run* (level dimension) *of the rafter,* the *run of the overhang,* or the *run between birdsmouths* by the *rake multiplier,* you can find the length of the rafter, the length of the overhang, and the distance between birdsmouths.

On the calculator this sequence looks like this:

B) Find the *rake multiplier.*
   (Unit rise) x^2 $+$ 12 x^2 $=$ $\sqrt{x}$ $\div$ 12 $=$ **(rake multiplier)**

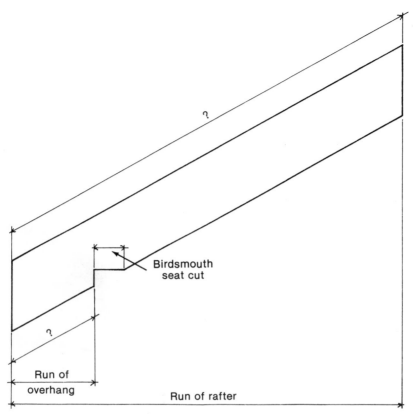

Draw the rafter first!
Figure 5-7

C) Now, find the length of the rafter.
   1) **(Run of the rafter)** ⊠ **(rake multiplier)** ⊟ **(length of the rafter in decimal inches)**
   2) Convert the length to fractional inches.

D) Next, find the length of the overhang (note that on the same roof or a roof with the same pitch, the *rake multiplier* will be the same).
   1) **(Run of overhang)** ⊠ **(rake multiplier)** ⊟ **(length of overhang in decimal inches)**
   2) Convert the length to fractional inches.

Common Rafter Layout 71

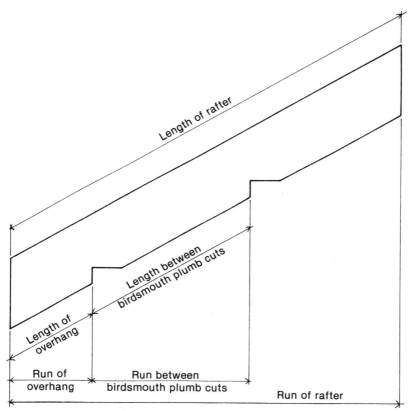

Finding rafter lengths
Figure 5-8

Once you know the length of the overhang, you know where to place the birdsmouth: you measure the length of the overhang from the end of the rafter to find where to put the plumb cut of the birdsmouth (Figure 5-8).

If there is more than one birdsmouth, you need to know the *run between birdsmouth plumb cuts* (Figure 5-8).

E) Last, find the distance between birdsmouth plumb cuts.
  1) **(Run between birdsmouth plumb cuts)** ☒ **(rake multiplier)** ☲ **(length between birdsmouth plumb cuts)**
  2) Convert the length to fractional inches.

72 Carpentry Layout

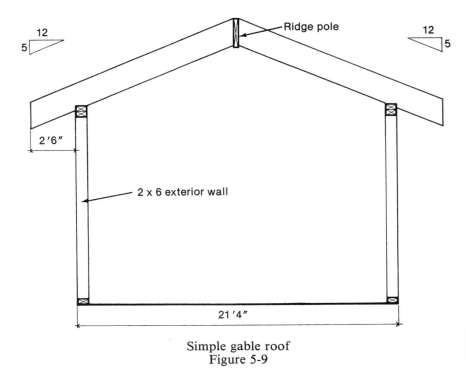

Simple gable roof
Figure 5-9

When there is more than one birdsmouth, figure the placement of the first birdsmouth and then measure from the plumb cut of that birdsmouth to locate the plumb cut of the next birdsmouth.

Applications

Here are some examples to illustrate these ideas.

Example 1
You are working on a building 21′4″ wide (Figure 5-9). The pitch of the roof is $5/12$ and the ridge pole is a 2 x 12 centered on the building. The rafter overhang is 2′6″. The rafter gets full bearing on a 2 × 4 exterior wall. Find all the dimensions necessary to lay out the rafter.
First draw the rafter and put in the dimensions you know (Figure 5-10):
- The *unit rise* is 5 (this is the first number in the pitch).
- The *run of the overhang* is 2′6″, or 30″.

Common Rafter Layout **73**

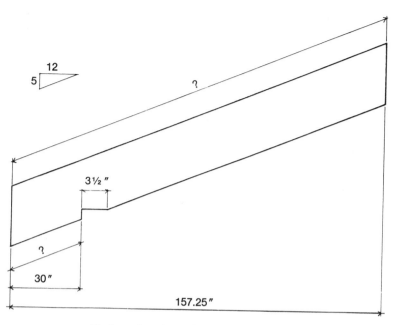

Rafter drawing with run dimensions
Figure 5-10

- The *run of rafter* is figured as follows.

A) Find the run of the rafter.
1) **(Ft.)** ⊠ 12 ⊞ **(in.)** ⊟ **(width of the building)**
   21 ⊠ 12 ⊞ 4 ⊟ 256
2) **(Width)** ⊟ 2 ⊟ **(half of width)**
   256 ⊟ 2 ⊟ 128
3) **(Half of building width)** ⊟ **(ridge pole)** ⊟ 2 ⊟ **(run of rafter inside the building)**
   128 ⊟ 1.5 ⊟ 2 ⊟ 127.25
4) **(Run of rafter inside building)** ⊞ **(run of overhang)** ⊟ **(run of rafter)**
   127.25 ⊞ 30 ⊟ 157.25

- The *run between birdsmouth plumb cuts* is not applicable in this case. Because the rafter goes from ridge pole to exterior wall, there is only one bearing point.

74 *Carpentry Layout*

- The *seat cut of the birdsmouth* is given in the dimension of the exterior wall: 2 × 4. The actual dimension of a nominal 4″ board is 3½″, so the *seat cut of the birdsmouth* here is 3.5″.

Now that you have gathered the information you need, you are ready to calculate.

B) Find, the *rake multiplier* for a $\frac{5}{12}$ roof pitch.
  (Unit rise) $\boxed{x^2}$ $\boxed{+}$ 12 $\boxed{x^2}$ $\boxed{=}$ $\boxed{\sqrt{x}}$ $\boxed{\div}$ 12 $\boxed{=}$ **(rake multiplier)**
  5     $\boxed{x^2}$ $\boxed{+}$ 12 $\boxed{x^2}$ $\boxed{=}$ $\boxed{\sqrt{x}}$ $\boxed{\div}$ 12 $\boxed{=}$ 1.0833333

C) Now, find the length of the rafter.
  1) (Run of the rafter) $\boxed{\times}$ **(rake multiplier)** $\boxed{=}$ **(length of rafter)**
        157.25 $\boxed{\times}$ 1.0833333 $\boxed{=}$ 170.35417
  2) Convert the length to fractional inches: 170.35417″ = 170⅜″.

D) Finally, find the length of the overhang.
  1) (Run of the overhang) $\boxed{\times}$ **(rake multiplier)** $\boxed{=}$ **(length of overhang)**
          30            $\boxed{\times}$ 1.0833333 $\boxed{=}$ 32.499999
  2) Convert the length to fractional inches: 32.499999″ = 32½″.

E) Since there is only one birdsmouth, there is no length between birdsmouth plumb cuts.

Write in these dimensions on your drawing, and hand the drawing to the saw man (see Figure 5-11).

Example 2
You must lay out the common rafters for a building 28′3¼″ wide. The ridge pole is a 1 × 12, and it is centered on the building (Figure 5-12). The rafters bear fully on the 2 × 6 exterior wall and the 2 × 4 interior bearing wall. The outside of the 2 × 4 wall is 10′2½″ from the outside of the 2 × 6 wall. The overhang is 2′ and the pitch of the roof is 7½/12. Find all the dimensions you need to lay out the common rafters.

First draw the rafter and put in the dimensions you know (Figure 5-13).
- The unit rise of the roof is 7½. This is given as the first number in the pitch, 7½/12.
- The run of the overhang is 2′, or 24″.
- The run of the rafters is found by figuring the level distance the rafter must go.

Common Rafter Layout 75

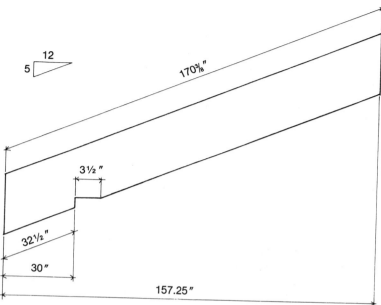

Rafter drawing with length dimensions
Figure 5-11

A) Find the run of the rafters.
   1) **(Ft.)** ⊠ 12 ⊞ **(in.)** ⊞ **(numerator)** ⊟ **(denominator)** ⊟ **(width of the building)**
      28 ⊠ 12 ⊞ 3 ⊞ 1 ⊟ 4 ⊟ 339.25
   2) **(Width)** ⊟ 2 ⊟ **(half of width)**
      339.25 ⊟ 2 ⊟ 169.625
   3) **(Half of building width)** ⊟ **(ridge pole)** ⊟ 2 ⊟ **(run of rafter inside the building)**
      169.625 ⊟ .75 ⊟ 2 ⊟ 169.25
   4) **(Run of rafter inside the building)** ⊞ **(run of overhang)** ⊟ **(run of the rafter)**
      169.25 ⊞ 24 ⊟ 193.25

- The run between birdsmouth plumb cuts is the distance from the outside of one bearing wall to the outside of the next; you are given this as 10′2½″, or 122 ½″.

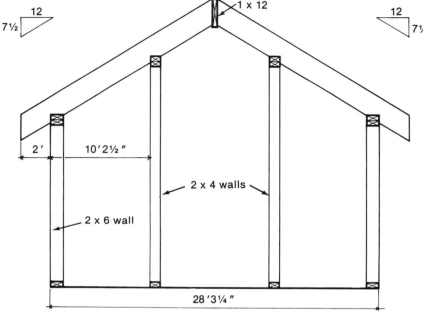

Gable roof with four bearing walls
Figure 5-12

- The seat cut required at the birdsmouths is the same as the thickness of the wall, since the rafters are fully bearing. The exterior wall is 2 × 6, so the *exterior seat cut* is 5½". The interior wall is 2 × 4, so the *interior seat cut* is 3½".

Put these dimensions on your drawing and you are ready to calculate.

B) Find the rake multiplier for a 7½/12 roof.
   1) (Unit rise) $\boxed{x^2}$ $\boxed{+}$ 12 $\boxed{x^2}$ $\boxed{=}$ $\boxed{\sqrt{x}}$ $\boxed{\div}$ 12 $\boxed{=}$ **(rake multiplier)**
      7.5 $\boxed{x^2}$ $\boxed{+}$ 12 $\boxed{x^2}$ $\boxed{=}$ $\boxed{\sqrt{x}}$ $\boxed{\div}$ 12 $\boxed{=}$ 1.1792476

Remember, multiply each run by the *rake multiplier* to get the lengths of the rafter.

C) Find the length of the rafter.

Common Rafter Layout 77

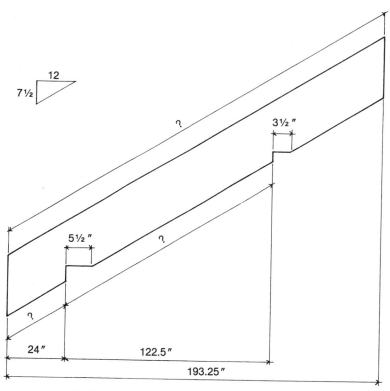

Rafter drawing with run dimensions
Figure 5-13

1) **(Run of the rafter)** ☒ **(rake multiplier)** ☐ **(length of rafter)**
   193.25   ☒   1.1792476   ☐   227.8896
2) Convert to fractional inches: 227.8896″ = 227⅞″.

D) Now find the overhang length.
   1) **(Run of the overhang)** ☒ **(rake multiplier)** ☐ **(length of overhang)**
         24   ☒   1.1792476   ☐   28.301942
   2) Convert the length to fractional inches: 28.301942″ = 28⁵⁄₁₆″.

E) Find the length between the plumb cuts of the birdsmouths.
   1) **(Run between birdsmouth plumb cuts)** ☒ **(rake multiplier)** ☐ **(length between birdsmouth plumb cuts)**
      122.5 ☒ 1.1792476 ☐ 144.45783

78 Carpentry Layout

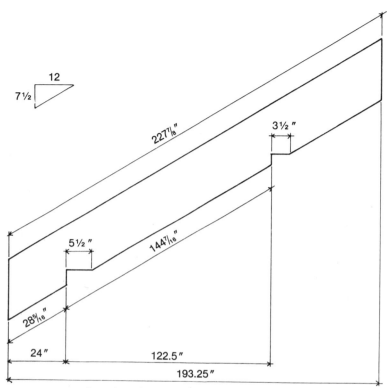

Rafter drawing with length dimensions
Figure 5-14

2) Convert the length to fractional inches: 144.45783″ = 144⁷⁄₁₆″.

Write in these dimensions on your drawing and you are ready to cut the rafters (see Figure 5-14).

Example 3

Now you are working on a building with glulam purlin beams (see Figure 5-15). There are three bearing points: the exterior wall, the ridge purlin, and the intermediate purlin. The outside face of the intermediate purlin is 12′3⅛″ from the outside of the exterior 2 × 4 bearing wall. The ridge purlin is centered on the building, which is 50′2¼″ wide. The purlins are 6¾″ × 21″ glulam beams, and the roof pitch is ⁸⁄₁₂. The rafters get full bearing on the 2 × 4 wall and 3⅜″ bear-

Common Rafter Layout 79

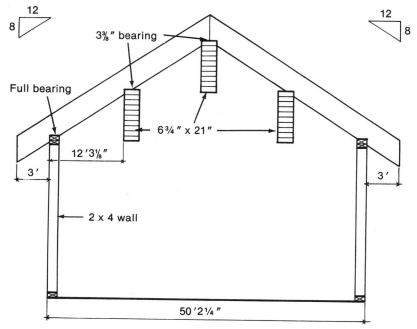

Roof framing with glulam purlins
Figure 5-15

ing on the ridge purlin, since you can only use half of the ridge purlin for the rafters on each side. The drawings stipulate that the rafters bear on the intermediate purlin 3⅜". The overhang is 3'. Find all the dimensions necessary to lay out the rafters.

First draw the rafter and write in the dimensions that you know (Figure 5-16).

- The *unit rise* is 8.
- The *run of the overhang* is 3', or 36".
- The *run of the rafter* is figured as follows.

A) There is no ridge pole (the rafters bear on the ridge purlin), so the run of the rafter is the run of the overhang (36") plus the distance from the exterior of the building to the center of the ridge purlin. That distance is half the width of the building in decimal inches and is figured as follows:

1) (Ft.) ☒ **12** ⊞ **(in.)** ⊞ **(numerator)** ⊡ **(denominator)** ⊟ **(width of building)**
50 ☒ 12 ⊞ 2 ⊞ 1 ⊡ 4 ⊟ 602.25

80 Carpentry Layout

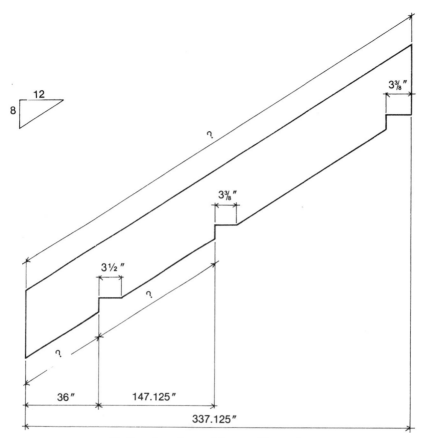

Rafter framing with run dimensions
Figure 5-16

2) **(Width)** ÷ 2 = **(half of width)**
   602.25 ÷ 2 = 301.125
3) **(Half of building width)** + **(run of overhang)** = **(run of rafter)**
   301.125   +   36   =   337.125

- The *run between the exterior and interior birdsmouth plumb cuts is* 12′3⅛″, or 147.125″.
- The *seat cut* required at the exterior birdsmouth is 3½″. At the other two birdsmouths, it is 3⅜″.

Put these dimensions on your drawing and you are ready to calculate.

B) Find the rake multiplier for an 8/12 roof.
 (Unit rise) x^2 + 12 x^2 = $\sqrt{x}$ ÷ 12 = **(rake multiplier)**
   8   x^2 + 12 x^2 = $\sqrt{x}$ ÷ 12 =   1.2018504

C) Now find the length of the rafter.
  1) (Run of the rafter) × **(rake multiplier)** = **(length of rafter)**
       337.125   ×   1.2018504   =   405.17381
  2) Convert the length to fractional inches: 405.17381″ = 405 3/16 ″.

D) Now, find the overhang length.
  1) (Run of the overhang) × **(rake multiplier)** = **(length of overhang)**
        36   ×   1.2018504   =   43.266614
  2) Convert to fractional inches: 43.266614″ = 43 1/4 ″.

E) Next, find the length between the plumb cuts of the exterior and interior birdsmouth.
  1) (Run between birdsmouth plumb cuts) × **(rake multiplier)** = **(length between birdsmouth plumb cuts)**
   147.125 × 1.2018504 = 176.82224
  2) Convert the length to fractional inches: 176.82224″ = 176 13/16 ″.

The birdsmouth at the ridge purlin can be found by setting the rafter square on an 8/12 pitch at the end of the rafter and measuring 3 3/8 ″ along the seat cut.

Write the dimensions you have calculated on the rafter drawing. You are ready to cut the rafter (Figure 5-17).

Example 4
You have taken on a job to build a house with a shed roof (Figure 5-18). The pitch is 1/12, and the width of the structure is 22′6″. There is a bearing wall at 11′1 1/4 ″ from the west side, and all the walls are 2 × 4s. The overhang is 2′2″ on each end. The rafters get full bearing. Find all the dimensions necessary to lay out the rafters.

Begin by drawing a picture and writing in the known dimensions (Figure 5-19).
- The unit rise is 1.
- The run of the overhang is 2′2″, or 26″.
- Now, find the run of the rafter.

A) Since this is not a gable roof, the *run of the rafter* is the width of the

82 *Carpentry Layout*

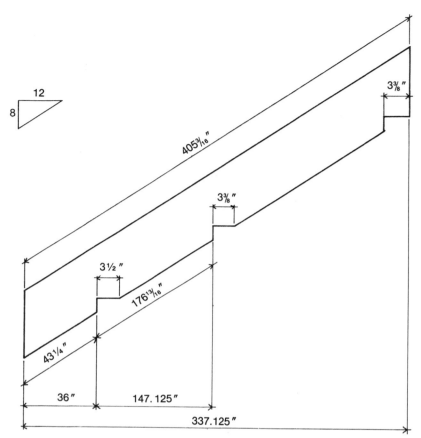

Rafter with length dimensions
Figure 5-17

building, in decimal inches, plus both overhangs in decimal inches.
(**Ft.**) ⊠ 12 ⊞ (**in.**) ⊞ (**overhang**) ⊞ (**overhang**) ▣ (**run of rafter**)
22 ⊠ 12 ⊞ 6 ⊞   26   ⊞   26   ▣   322

- The run between the exterior and interior birdsmouth plumb cuts is 11′1¼″, or 133.25″. The run between the two exterior birdsmouth plumb cuts is equal to the width of the building less the thickness of the wall (3½″) on which the exterior birdsmouth bears. On the calculator:
270 ⊟ 3.5 = 266.5

Common Rafter Layout

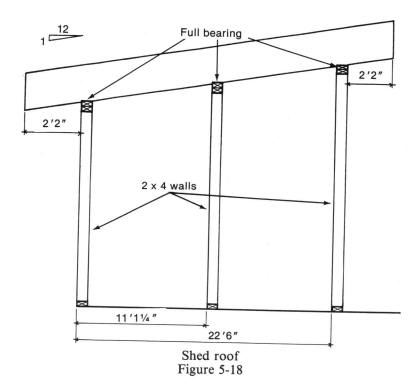

Shed roof
Figure 5-18

- Since all the bearing walls are 2 × 4 walls, all the birdsmouth seat cuts are 3.5″.

Now we are ready to calculate the remaining dimensions.

B) Find the rake multiplier for a $\frac{1}{12}$ pitch.
   (Unit rise) x^2 $+$ 12 x^2 $=$ $\sqrt{x}$ $\div$ 12 $=$ **(rake multiplier)**
   1   x^2 $+$ 12 x^2 $=$ $\sqrt{x}$ $\div$ 12 $=$ 1.0034662

C) Next, figure the length of the rafter.
   1) (Run of the rafter) $\times$ **(rake multiplier)** $=$ **(length of rafter)**
          322         $\times$ 1.0034662 $=$ 323.11612
   2) Convert to fractional inches: 323.11612″ = 323$\frac{1}{8}$″.

D) Next, find the length of the overhang.
   1) (Run of overhang) $\times$ **(rake multiplier)** $=$ **(length of overhang)**
          26          $\times$ 1.0034662 $=$ 26.090122

2) Convert the length to fractional inches: $26.090122'' = 26\frac{1}{16}''$.

E) You have two different calculations to do for birdsmouth plumb cuts. First, figure the length between the plumb cuts of the exterior and interior birdsmouths.
  1) **(Run between birdsmouth plumb cuts)** ⊠ **(rake multiplier)** ⊟ **(distance between exterior and interior plumb cuts)**
     133.25 ⊠ 1.0034662 ⊟ 133.71187
  2) Convert the length to fractional inches: $133.71187'' = 133\frac{11}{16}''$.

Finally, find the length between the two exterior birdsmouth plumb cuts.
  1) **(Run between birdsmouth plumb cuts)** ⊠ **(rake multiplier)** ⊟ **(length between exterior plumb cuts)**
     266.5 ⊠ 1.0034662 ⊟ 267.42374
  2) Convert the length to fractional inches: $267.42374'' = 267\frac{7}{16}''$.

Write all these dimensions in on the rafter drawing and you are ready to cut (Figure 5-20).

Example 5
You are figuring the common rafter layout for a structure with a $\frac{10}{12}$ pitch roof (Figure 5-21). The rafters get full bearing at the 2 × 6 exterior walls and on the $8\frac{3}{4}''$ × $21''$ glulam ridge purlin. The building is 24' wide, and the ridge purlin is centered at $7'2''$ from the west side of the house. The overhang on each side is 2'. Find all the layout dimensions for the common rafters.

First, draw the rafters and put in the dimensions you know (Figure 5-22).
- The unit rise is 10.
- The run of the overhang is $24''$.
- There is no ridge pole (the rafters bear on the ridge purlin), so the run of the rafter is the run of the overhang (2') plus the distance from the exterior of the building to the center of the ridge purlin. On the west side that distance is $7'2''$.

A) Find the run of the rafter on the west side.
  1) Convert the west side distance to inches.
     (Ft.) ⊠ 12 ⊞ **(in.)** ⊟ **(distance)**
     7   ⊠ 12 ⊞ 2   ⊟   86

Common Rafter Layout **85**

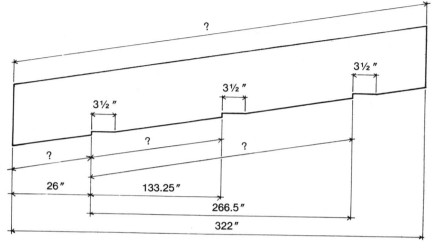

Rafter with run dimensions
Figure 5-19

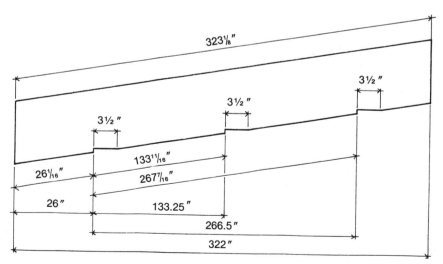

Rafter with length dimensions
Figure 5-20

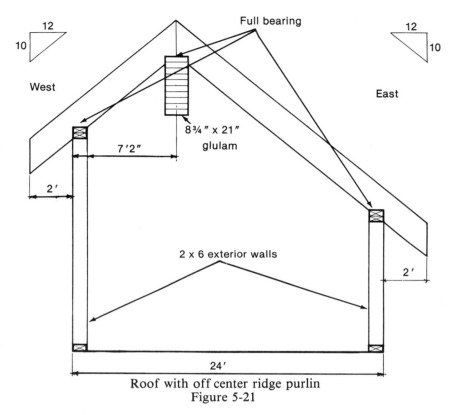

Roof with off center ridge purlin
Figure 5-21

2) Add the run of the overhang.
   (Distance) ⊞ **(run of overhang)** ⊟ **(run of rafter)**
       86   ⊞   24   ⊟ 110

B) Find the run of the rafter on the east side.
   1) Convert the building width to inches.
      (Ft.) ⊠ **12** ⊞ **(in.)** ⊟ **(width)**
       24 ⊠ **12** ⊞ 0 ⊟ **288**
   2) Subtract west side distance from the width of the building.
      288 ⊟ 86 ⊟ 202
   3) Add the run of the overhang, and you will have the *run of the rafter* on the east side.
      202 ⊞ 24 ⊟ 226

- The run between birdsmouth plumb cuts is not required. You will find the placement of the exterior birdsmouth from the overhang length. You can place the ridge purlin birdsmouth by using a rafter square since it will be at the end of the rafter.
- The *seat cut* of the first birdsmouth is 5½", or 5.5" (full bearing

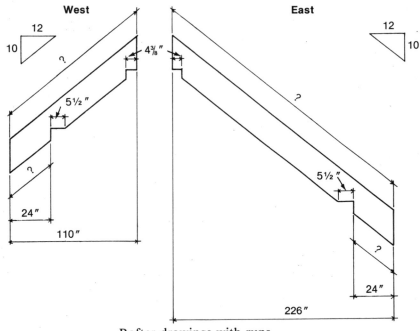

Rafter drawings with runs
Figure 5-22

on a 2 × 6 wall). The *seat cut* of the ridge purlin birdsmouth is half the thickness of the beam, 4⅜" or 4.375".

Put these dimensions on your drawing and you are ready to calculate.

C) First, find the rake multiplier for a ¹⁰⁄₁₂ pitch roof.
    (Unit rise) x^2 + 12 x^2 = $\sqrt{x}$ ÷ 12 = **(rake multiplier)**
    10   x^2 + 12 x^2 = $\sqrt{x}$ ÷ 12 = 1.3017083

Next, figure the rafter layout for the common rafters on the *west side*. Multiply each run by the rake multiplier to find the rafter length and birdsmouth placement. Start by finding the full rafter length.

D) Find the length of the rafter.
    1) (Run of the rafter) × **(rake multiplier)** = **(length of rafter)**
            110   ×   1.3017083   =   143.18791

88 *Carpentry Layout*

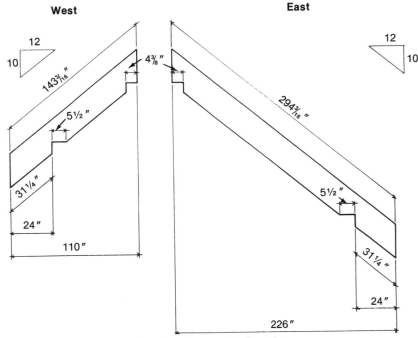

Rafter drawings with lengths
Figure 5-23

2) Convert the length to fractional inches: 143.18791" = 143³⁄₁₆".

E) Next, find the length of the overhang.
  1) **(Run of the overhang)** ⊠ **(rake multiplier)** ⊟ **(length of overhang)**
      24   ⊠   1.3017083   ⊟   31.240999
  2) Convert the length to fractional inches: 31.240999" = 31¼".

The birdsmouth at the ridge purlin is placed by setting the rafter square for a ¹⁰⁄₁₂ pitch at the end of the rafter and measuring back 4⅜" along the seat cut (Figure 5-23).
  Now, figure the layout for a common rafter on the *east side*.

D) First, calculate the full length of the rafter.
  1) **(Run of the rafter)** ⊠ **(rake multiplier)** ⊟ **(length of rafter)**
      226   ⊠   1.3017083   ⊟   294.18607
  2) Convert the length to fractional inches: 294.18607" = 294 ³⁄₁₆".

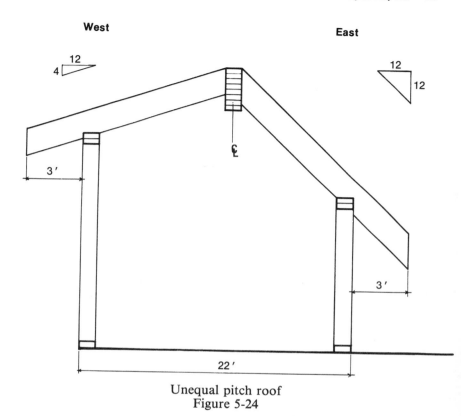

Unequal pitch roof
Figure 5-24

E) Finally, find the overhang length.
  1) **(Run of the overhang)** ☒ **(rake multiplier)** ☐ **(length of overhang)**
       24   ☒   1.3017083   ☐   31.240999
  2) Convert the length to fractional inches: 31.240999" = 31¼".

The birdsmouth at the ridge purlin is the same as for the common rafter on the west side.

Write in these dimensions on your rafter drawing and you are ready to cut the rafters (Figure 5-23).

Example 6
You are framing a gable wall in a structure with an unequal-pitch roof (Figure 5-24). The west side is $4/12$ and the east is $12/12$. The ridge pole is a 5⅛" × 16½" glulam beam centered on the building. The exterior

90 *Carpentry Layout*

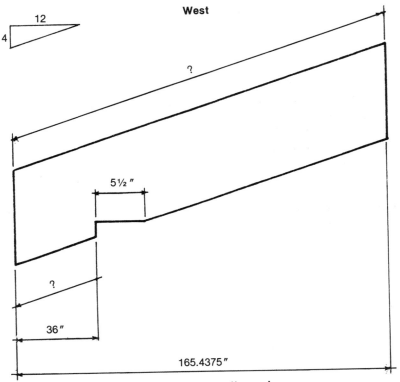

West rafter with run dimensions
Figure 5-25

bearing walls are 2 × 6s and the overhang is 3′ on each side. The rafters get full bearing and the building is 22′ wide. Figure all the dimensions necessary to lay out the rafters.

Start with the west rafters. First, draw a rafter and put in the dimensions you know (Figure 5-25).
- The unit rise is 4.
- The run of the overhang is 3′, or 36″.
- The run of the rafter is the level distance the rafter must cover.

A) Find the run of the rafter.
   1) Convert the building width to decimal inches.
      (Ft.) ☒ 12 ☒ **(building width)**
      22   ☒ 12 ☒   264

2) Divide the width of the building in half (because the ridge pole is centered).
(Building width) ÷ 2 = **(half of width)**
   264   ÷ 2 = 132
3) Divide the width of the ridge pole in half and subtract it from half the building width.
(Half of building width) − **(ridge pole)** ÷ 2 = **(run of rafter inside the building)**
   132 − 5.125 ÷ 2 = 129.4375
4) Add the run of the overhang.
(Run of rafter inside building) + **(run of overhang)** = **(run of rafter)**
   129.4375   +   36   =   165.4375

- Since there is only one birdsmouth, the *run between birdsmouth plumb cuts* is not applicable.
- The *seat cut* required at the birdsmouth is 5½″, or 5.5″, since the rafters get full bearing.

After writing in these dimensions on your drawing, you are ready to calculate.

B) First, find the rake multiplier for a 4/12 roof.
(Unit rise) x^2 + 12 x^2 = $\sqrt{x}$ ÷ 12 = **(rake multiplier)**
   4   x^2 + 12 x^2 = $\sqrt{x}$ ÷ 12 = 1.0540926

Now multiply each run by the *rake multiplier* to get the lengths of the rafter.

C) Find the length of the rafter.
  1) **(Run of the rafter)** × **(rake multiplier)** = **(length of rafter)**
       165.4375   ×   1.0540926   =   174.38644
  2) Convert the length to fractional inches: 174.38644″ = 174⅜″.

D) Next, find the length of the overhang.
  1) **(Run of the overhang)** × **(rake multiplier)** = **(length of overhang)**
        36   ×   1.0540926   =   37.947333
  2) Convert the length to fractional inches: 37.947333″ = 37¹⁵⁄₁₆″.

Next, figure the layout for the rafters on the *east side*. Draw a rafter and put in the dimensions you know in decimal inches (Figure 5-26).

92 *Carpentry Layout*

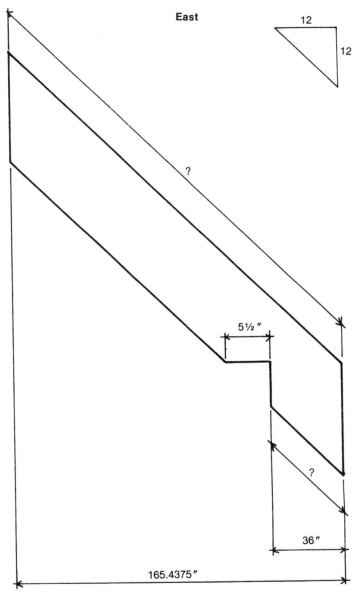

East rafter with run dimensions
Figure 5-26

- The unit rise is 12.
- The run of the overhang is 36″.
- The run of the rafter is the same as on the west side: 165.4375″.
- Again, since there is only one birdsmouth, there is no run between birdsmouth plumb cuts.
- The seat cut required at the birdsmouth is 5.5″, since the rafter gets full bearing.

Write these dimensions on the rafter drawing and you are again ready to calculate.

B) First, find the rake multiplier for a 12/12 roof.
   (Unit rise) x^2 $+$ 12 x^2 $=$ $\sqrt{x}$ $\div$ 12 $=$ **(rake multiplier)**
   12 $\quad$ x^2 $+$ 12 x^2 $=$ $\sqrt{x}$ $\div$ 12 $=$ 1.4142136

Now, multiply each run by the *rake multiplier* to get the lengths of the rafter.

C) Start by finding the length of the rafter.
   1) **(Run of the rafter)** $\times$ **(rake multiplier)** $=$ **(length of rafter)**
      165.4375 $\quad\times\quad$ 1.4142136 $\quad=\quad$ 233.96396
   2) Convert to fractional inches: 233.96396″ = 233 15/16″.

D) Next find the length of the overhang.
   1) **(Run of the overhang)** $\times$ **(rake multiplier)** $=$ **(length of the overhang)**
      36 $\quad\times\quad$ 1.4142136 $\quad=\quad$ 50.911689
   2) Convert to fractional inches: 50.911689″ = 50 15/16″.

Write in all these dimensions on your rafter drawings and you are ready to cut the rafters (Figures 5-27 and 5-28)

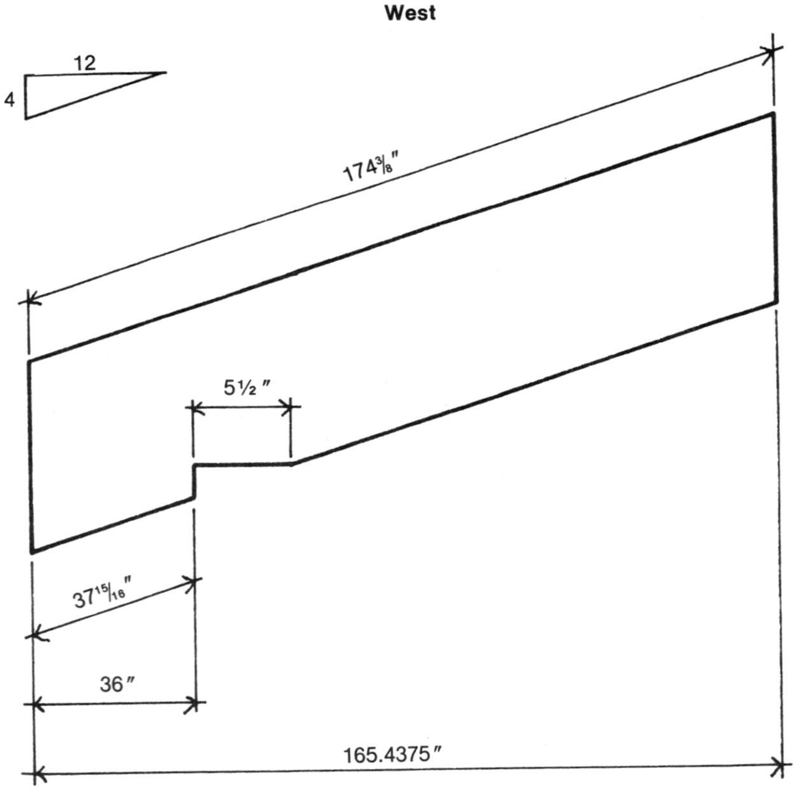

94 *Carpentry Layout*

West

West rafter with length dimensions
Figure 5-27

Common Rafter Layout 95

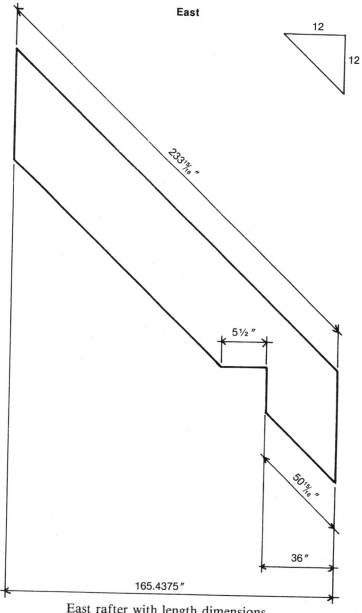

East rafter with length dimensions
Figure 5-28

Problems

1) You are working on a building 18′1¼″ wide. The roof pitch is 10/12 and the ridge pole is a 1 × 12 centered on the building. The overhang is 2′ and the rafters get full bearing on a 2 × 6 wall. Find all the dimensions necessary to lay out the rafter.

Hint: On each problem, draw the building and the rafter first.

2) You must lay out the rafters for a building 41′ wide. The ridge pole is a 2 × 12 centered on the building. The rafters bear fully on the 2 × 4 walls. The outside of the exterior bearing wall is 11′3″ from the outside of the interior bearing wall. The overhang is 8″ and the roof pitch is 9¼/12. Find all the dimensions you need to lay out the common rafters.

3) You are working on a building with glulam purlin beams 5⅛″ × 15″ at three bearing points: the exterior wall, the ridge line, and an intermediate purlin. The building is 31′1½″ with the ridge purlin centered on it. The outside face of the intermediate purlin is 7′6″ from the exterior purlin. The roof pitch is 2/12, the overhang is 18″, and the rafters get full bearing. Find all the dimensions necessary to lay out the rafters.

Worksheet

COMMON RAFTER LAYOUT

Draw the rafter and put in the dimensions you know. This procedure works for a gable roof.

A) First, find the run of the rafter.
   1) Convert the building width to decimal inches.
      (Ft.) $\times$ 12 $+$ **(in.)** $+$ **(numerator)** $\div$ **(denominator)** $=$ **(building width)**
      _____ $\times$ 12 $+$ _____ $+$ _____ $\div$ _____ $=$ _____
   2) Divide the width of the building in half.
      (Building width) $\div$ 2 $=$ **(half of width)**
      _____ $\div$ 2 $=$ _____
   3) Divide the width of the ridge pole in half and subtract it from half the building width.
      (Half of building width) $-$ **(ridge pole)** $\div$ 2 $=$ **(run of the rafter inside the building)**
      _____ $-$ _____ $\div$ 2 $=$ _____
   4) Add the run of the overhang.
      (Run of the rafter inside building) $+$ **(run of the overhang)** $=$ **(run of the rafter)**
      _____ $+$ _____ $=$ _____

B) Now, find the rake multiplier.
   (Unit rise) x^2 $+$ 12 x^2 $=$ $\sqrt{x}$ $\div$ 12 $=$ **(rake multiplier)**
   _____ x^2 $+$ 12 x^2 $=$ $\sqrt{x}$ $\div$ 12 $=$ _____

C) Next, find the length of the rafter.
   1) **(Run of the rafter)** $\times$ **(rake multiplier)** $=$ **(length of the rafter)**
      _____ $\times$ _____ $=$ _____
   2) Convert the length to fractional inches.

D) Next, find the length of the overhang.
   1) **(Run of the overhang)** $\times$ **(rake multiplier)** $=$ **(length of the overhang)**
      _____ $\times$ _____ $=$ _____
   2) Convert the length to fractional inches.

E) Last, find the length between birdsmouth plumb cuts.
   1) (Run between birdsmouth plumb cuts) ☒ **(rake multiplier)** ☐ **(length between birdsmouth plumb cuts)**
   _____ ☒ _____ ☐ _____
   2) Convert the length to fractional inches.

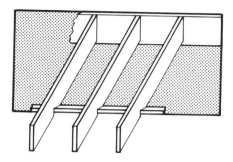

6
Rafter Layout in the Field

Suppose you have calculated a common rafter as in Example 1 of Chapter 5. It is shown in Figure 5-11. Now you must cut it. What do you do next?

Cutting Rafters

To cut this rafter, you will need the following:
- A rafter square with markers.
- A tape measure at least 16' long.
- A rafter drawing, as in Figure 5-11.

First choose a piece of lumber for the rafter that is 16' long. Crown the piece by sighting along the edge of it and drawing an arrow pointing to the hump. This arrow points to the top of the rafter (Figure 6-1). Place the rafter on a pair of saw horses with the crown away from you. This will help you visualize the rafter in place and will reduce the chance for error.

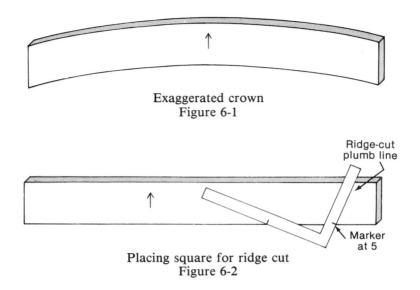

Exaggerated crown
Figure 6-1

Placing square for ridge cut
Figure 6-2

Set the markers at 5 on the tongue and 12 on the body of the square. Place the square on the rafter stock as shown in Figure 6-2 and draw the ridge cut along the tongue. All lines drawn along the tongue of the square will be plumb, while those drawn along the body will be level.

THE TAIL CUT

Now measure along the bottom edge (the edge away from the direction the arrow points) of the rafter a distance of 170 3/8" from the ridge line, and mark that point (Figure 6-3). Slide the rafter square to that mark and line the tongue of the square up with it (Figure 6-4). Draw a line along the tongue: That is the tail cut.

THE BIRDSMOUTH CUT

Next measure along the bottom edge of the rafter 32 1/2" from the tail cut and mark that point. Slide the rafter square until the tongue lines up with this mark. Draw the birdsmouth plumb cut (Figure 6-5).

To finish off the birdsmouth, you must draw a seat cut 3 1/2" long. Since this line will be level, you will draw it along the body of the rafter square. The marker on the body is set at 12", so to get a 3 1/2"

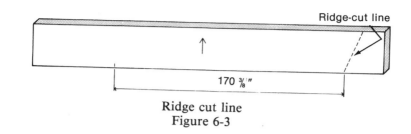

Ridge cut line
Figure 6-3

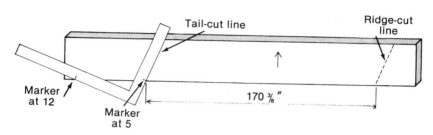

Placing square for tail cut
Figure 6-4

seat-cut line, you must line up the birdsmouth plumb-cut line with 15½" on the body. When this is lined up, draw in the birdsmouth seat cut (Figure 6-6).

Cutting the rafter is straightforward: Cut along the lines and finish the birdsmouth with a handsaw or chisel.

Fitting the Rafters

Try this rafter in several places on the building. If it fits everywhere, use it as a pattern and cut all the rafters from it. If it doesn't fit in some places, make a note of how far off it is, then adjust the rafters in those places accordingly. This will happen if the building is slightly out of square or plumb.

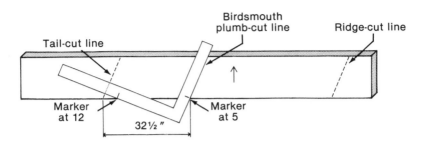

Placing square for birdsmouth plumb cut
Figure 6-5

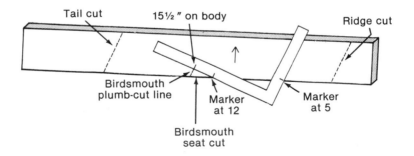

Placing square for birdsmouth seat cut
Figure 6-6

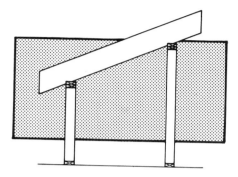

7
Bearing-Wall Heights

When there is an interior bearing wall, it is critical that the wall be the correct height so that the rafters actually do bear on it. In this chapter you learn a method of calculating the exact height of bearing walls or beams for any roof pitch.

In the past you may have unintentionally changed the roof pitch of a building by framing the walls at the wrong height. Or, you may have framed a series of three bearing walls in such a way that the rafters wouldn't bear on all the walls at once. If you use the method outlined in this chapter, these mistakes should all be in the past.

Calculating the Heights

In order to calculate the bearing-wall heights, you need to know:
- The *roof pitch*.
- The *amount of rafter bearing* at each wall (Figure 7-1).
- The *exterior bearing-wall height* (Figure 7-1).

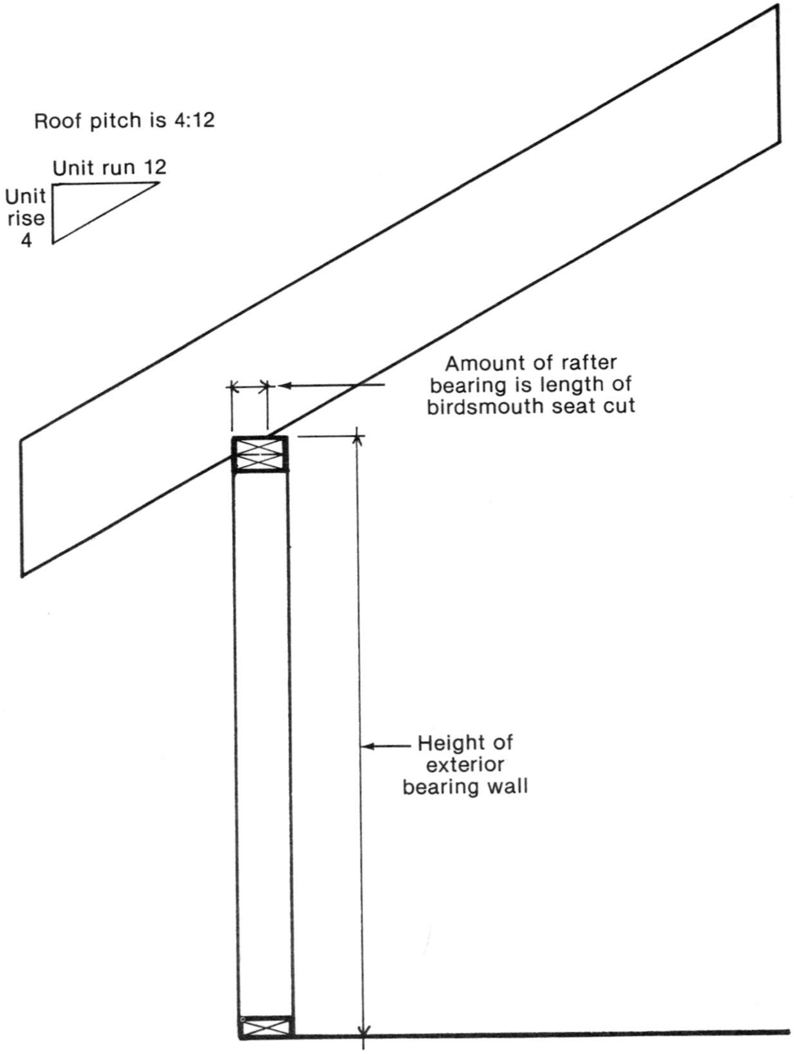

104 *Carpentry Layout*

Roof pitch is 4:12

Unit run 12
Unit rise 4

Amount of rafter bearing is length of birdsmouth seat cut

Height of exterior bearing wall

Dimensions needed for framing interior bearing walls
Figure 7-1

TERMS

Roof Pitch. The *roof pitch* is usually given on the plans as a fraction or proportion; for example, 4/12 or 4:12 may be given, as in Figure 7-1. In this example, the roof rises 4 units (inches, feet, meters) for every 12 of the same units (inches, feet, meters) of run; you will recall that the first number is the *unit rise* and the second number is the *unit run*. The unit run is always 12, with the *unit rise* varying to give different roof pitches.

Rafter Bearing. The **amount of rafter bearing** tells you the length of the seat cut at the birdsmouth (Figure 7-1). As you learned in Chapter 5, *full bearing* means the seat cut at the birdsmouth is the same length as the thickness of the bearing wall. Rafter bearing is always full bearing unless otherwise noted in the plans.

Exterior Bearing-wall Height. The **exterior bearing-wall height** is given in the blueprints on an elevation or a section. This is the distance from the bottom of the sole plate to the top of the double-top plate.

Beginning Work

When you know these three numbers, you are ready to calculate. First, you find the **support run.** This is the level distance from the outside of the exterior bearing wall to the near side of the interior bearing wall, less the amount of rafter bearing at the exterior bearing wall, plus the *amount of rafter bearing* at the interior bearing wall (Figure 7-2). Second, you find the **height multiplier** by dividing the *unit rise* by the *unit run*. Third, you multiply the support run by the *height multiplier* to find the added height. Finally, you add the *added height* to the *exterior-wall height* to obtain the interior wall height.

On a calculator, you proceed as follows.

A) Find the *support run*.
  (Distance to interior wall) ⊟ (rafter bearing at exterior wall) ⊞ (rafter bearing at interior wall) ⊟ (support run)

B) Find the *height multiplier*.
  (Unit rise) ⊟ (unit run) ⊟ (height multiplier)

C) Find the *added height*.
  (Height multiplier) ⊠ (support run) ⊟ (added height)

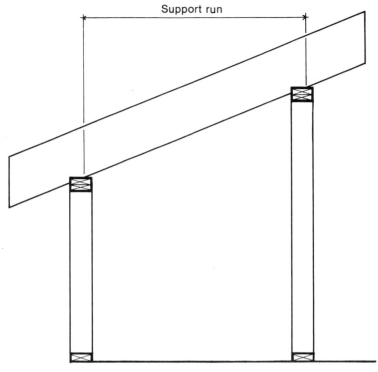

Finding the support run
Figure 7-2

D) Find the *interior-wall height*.
  1) **(Added height)** ⊞ **(exterior-wall height)** ⊟ **(interior-wall height)**
  2) Convert the height to fractional inches.

Applications

The following examples show you how to use this method. Watch for chances to use the display from one answer to calculate the next.

Example 1
You are framing an interior 2 x 4 bearing wall. The distance from the exterior of the building to the interior bearing wall is 102½" (Figure 7-3). The exterior bearing wall is a 2 × 4 wall 96⅞" high. The pitch of the roof is to be $\frac{5}{12}$ and the rafters are to have full bearing on both walls.

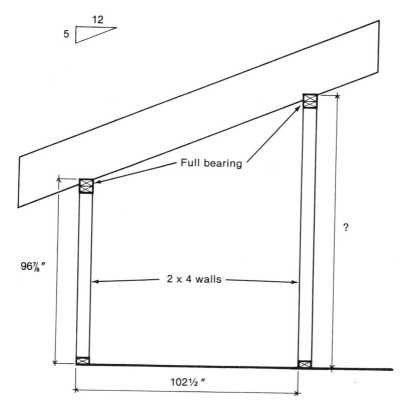

Finding the interior bearing-wall height
Figure 7-3

What is the height of the interior bearing wall?
   Start by listing the information you need.
- The roof pitch is ⁵⁄₁₂.
- The amount of rafter bearing is the same at both walls: 3½″, or 3.5″.
- The exterior bearing-wall height is 96⅞″.
   Now you are ready to calculate.

A) First, find the support run.
   (Distance to interior wall) ⊟ (rafter bearing at exterior wall) ⊞ (rafter bearing at interior wall) ⊟ (support run)
   102.5 ⊟ 3.5 ⊞ 3.5 ⊟ 102.5

B) Next, find the *height multiplier* for a ⁵⁄₁₂ pitch.
   (Unit rise) ÷ **(unit run)** = **(height multiplier)**
       5   ÷   12   =   .41666667

C) Then, find the *added height*.
   (Height multiplier) × **(support run)** = **(added height)**
       .41666667   ×   102.5   =   42.708333

D) Finally, figure the interior bearing-wall height.
   1) (Added height) + **(exterior-wall height)** = **(interior-wall height)**
        42.708333   +   96.875   =   139.58333
   2) Convert the height to fractional inches: 139.58333″ = 139⁹⁄₁₆″.

This is the height of the interior bearing wall.

Example 2
The building you are now framing has a 2 × 6 exterior bearing wall 96″ high (Figure 7-4). The rafters also bear on an 8¾″ × 21″ glulam beam. The level distance from the exterior of the building to the beam is 12′3⅛″. The rafters are to bear 4″ on both the 2 × 6 wall and the beam. The roof pitch is ¹⁰⁄₁₂. Find the height of the beam.
   Here is the information you know.
- The roof pitch is ¹⁰⁄₁₂.
- The amount of rafter bearing is 4″ at both exterior wall and beam.
- The exterior bearing-wall height is 96″.

You are now ready to calculate.

A) First, find the support run.
   (Distance to wall) − **(rafter bearing at exterior wall)** + **(rafter bearing at interior wall)** = **(support run)**
   147.125 − 4 + 4 = 147.125

B) Next, find the height multiplier for a ¹⁰⁄₁₂ pitch.
   (Rise) ÷ **(run)** = **(height multiplier)**
       10   ÷   12   =   .83333333

C) Then, figure the added height.
   (Height multiplier) × **(support run)** = **(added height)**
       .83333333   ×   147.125   =   122.60417

D) Finally, find the wall height.

Bearing-Wall Heights   109

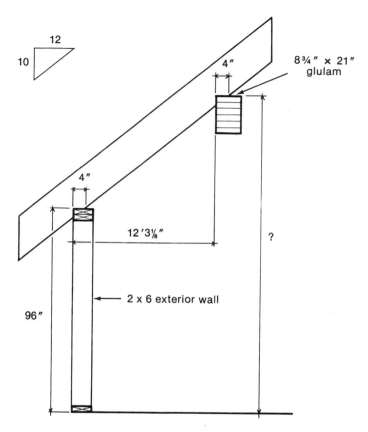

Finding the glulam height
Figure 7-4

1) (Added height) ⊞ **(exterior-wall height)** ⊟ **(interior-wall height)**
    122.60417   ⊞   96   ⊟   218.60417
2) Convert the height to fractional inches: 218.60417″ = 218⅝″.

The height of the interior bearing wall is 218⅝″.

Example 3
You are building a structure with three supporting walls on one side of the ridge: the exterior wall, which is 92″ high, and two interior walls. The exterior wall is a 2 × 6 wall, while the interior walls are 2 × 4 (Figure 7-5). The rafters are to be full bearing on all the walls. The first interior wall is 7′3¼″ from the exterior, and the second one is 13′9⅛″.

110 *Carpentry Layout*

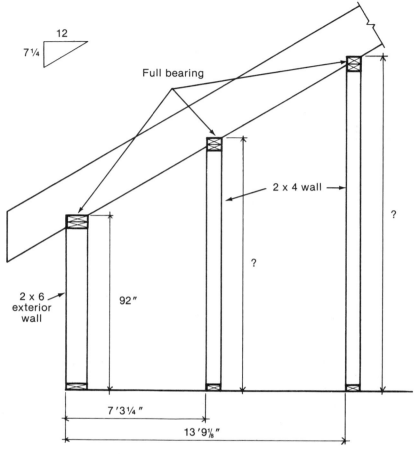

Figuring two interior bearing walls
Figure 7-5

If the pitch of the roof is 7¼/12, find the heights of the two interior walls.
   Summarize the information you know.
- The roof pitch is 7¼/12.
- The amount of rafter bearing at the exterior wall is 5½″; at the first interior wall, it is 3½″; and at the second, it is also 3½″.
- The exterior bearing-wall height is 92″.

Now, you are ready to calculate the heights of the interior walls. Begin with the first interior bearing wall.

A) First, find the support run for the first interior bearing wall.
(Distance to interior wall) ⊟ **(rafter bearing at exterior wall)** ⊞ **(rafter bearing at interior wall)** ⊟ **(support run)**
87.25 ⊟ 5.5 ⊞ 3.5 ⊟ 85.25

B) Next, calculate the height multiplier.
(Rise) ⊟ **(run)** ⊟ **(height multiplier)**
7.25 ⊟ 12 ⊟ .60416667

C) Now, find the added height for the first interior wall.
(Height multiplier) ⊠ **(support run)** ⊟ **(added height)**
.60416667 ⊠ 85.25 ⊟ 51.505208

D) Finally, figure the wall height for the first interior wall.
1) **(Added height)** ⊞ **(exterior-wall height)** ⊟ **(interior-wall height)**
51.505208 ⊞ 92 ⊟ 143.50521
2) Convert the height to fractional inches: 143.50521" = 143½".

Now, work on the second interior wall.

A) First, find the support run for the second interior bearing wall.
(Distance to wall) ⊟ **(rafter bearing at exterior wall)** ⊞ **(rafter bearing at interior wall)** ⊟ **(support run)**
165.125 ⊟ 5.5 ⊞ 3.5 ⊟ 163.125

B) Since you are still working on the same roof, with the same pitch, you can use the same height multiplier as for the first wall, .60416667.

C) Now, find the added height.
(Height multiplier) ⊠ **(support run)** ⊟ **(added height)**
.60416667 ⊠ 163.125 ⊟ 98.554688

D) Finally, figure the interior wall height.
1) **(Added height)** ⊞ **(exterior-wall height)** ⊟ **(interior-wall height)**
98.554688 ⊞ 92 ⊟ 190.554688
2) Convert the height to fractional inches: 190.554688" = 190 9/16".

Because this is a more complicated problem, summarize your information.
- Exterior-wall height = 92" (given)

112 *Carpentry Layout*

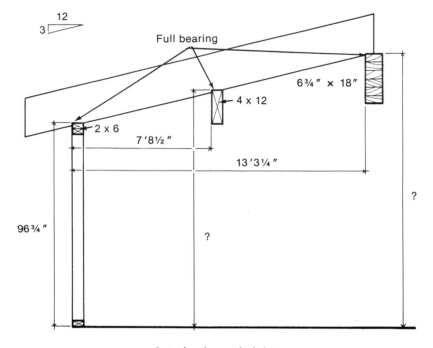

Interior beam heights
Figure 7-6

- First interior-wall height = 143½"
- Second interior-wall height = 190⁹⁄₁₆"

Example 4
The building on which you are working has a ³⁄₁₂ pitch roof (Figure 7-6). The exterior bearing wall is 2 × 6, there is to be an interior 4 × 12 bearing beam 7'8½" from the exterior, and the ridge support is a 6¾" × 18" glulam beam at 13'3¼" from the exterior. If the 2 × 6 exterior wall is 96¾" high and the rafters get full bearing, what are the heights of the beams?

Here is the information you know:
- The roof pitch is ³⁄₁₂.
- The amount of rafter bearing is 5½" at the exterior wall; at the 4 × 12 it is 3½"; and at the ridge support it is 3⅜" (half the width of the glulam beam) (Figure 7-6).
- The exterior bearing-wall height is 96¾".

First, calculate the height of the 4 × 12 (interior) beam.

A) Find the support run to the 4 × 12 beam.
(Distance to interior wall) ⊟ **(rafter bearing at exterior wall)** ⊞ **(rafter bearing at interior wall)** ⊟ **(support run)**
92.5 ⊟ 5.5 ⊞ 3.5 ⊟ 90.5

B) Next, find the height multiplier for a 3/12 pitch roof.
(Rise) ⊡ **(run)** ⊟ **(height multiplier)**
   3   ⊡   12   ⊟   .25

C) Then, find the added height for the 4 × 12 beam.
(Height multiplier) ⊠ **(support run)** ⊟ **(added height)**
     .25   ⊠   90.5   ⊟   22.625

D) Finally, calculate the 4 × 12 beam height.
  1) (Added height) ⊞ **(exterior-wall height)** ⊟ **(interior-wall height)**
        22.625   ⊞   96.75   ⊟   119.375
  2) Convert the height to fractional inches: 119.375″ = 119 3/8″.

Now, you need to work on the ridge-support-beam height.

A) Find the support run for the ridge-support beam.
(Distance to interior wall) ⊟ **(rafter bearing at exterior wall)** ⊞ **(rafter bearing at interior wall)** ⊟ **(support run)**
159.25 ⊟ 5.5 ⊞ 3.375 ⊟ 157.125

B) The height multiplier is the same as for the interior beam: .25.

C) Next, find the added height for the ridge-support beam.
(Height multiplier) ⊠ **(support run)** ⊟ **(added height)**
     .25   ⊠   157.125   ⊟   39.28125

D) Finally, find the beam height.
  1) (Added height) ⊞ **(exterior-wall height)** ⊟ **(interior-wall height)**
        39.28125   ⊞   96.75   ⊟   136.03125
  2) Convert the height to fractional inches: 136.03125″ = 136 1/16″.

Again, it'd be a good idea to summarize the heights before you begin work.
- Exterior-wall height = 96 3/4″ (given)
- 4 × 12 beam height = 119 3/8″
- Ridge- beam height = 136 1/16″

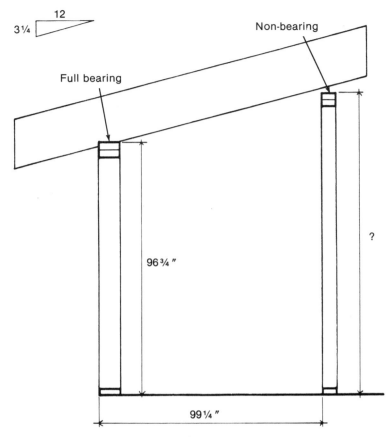

Finding the height of interior nonbearing partition
Figure 7-7

Example 5

Now that you are improving, you have a different job. You are framing a residence with a 3¼/12 roof (Figure 7-7). The only bearing walls are the exterior walls. There is a nonbearing partition 99¼" from the outside of the house, which is easier to stand before the roof is framed. The rafters are to get full bearing on the 2 × 4 exterior wall, which is 96¾" high. Since the interior partition is nonbearing, there will be no birdsmouth for it. Find the height of this interior partition.

Here is the information you know:
- The roof pitch is 3¼/12.
- The amount of rafter bearing is 3½" at the exterior bearing wall;

at the interior partition it is 0" (there will be no birdsmouth).
- The exterior-bearing-wall height is 96¾".

Now, you are ready to follow the four steps to find the wall height.

A) First, find the support run.
(Distance to interior wall) ⊟ **(rafter bearing at exterior wall)** ⊞ **(rafter bearing at interior wall)** ⊟ **(support run)**
99.25 ⊟ 3.5 ⊞ 0 ⊟ 95.75

B) Now, find the height multiplier for a 3¼/12 pitch.
(Rise) ⊡ **(run)** ⊟ **(height multiplier)**
3.25 ⊡ 12 ⊟ .27083333

C) Next, find the added height.
(Height multiplier) ⊠ **(support run)** ⊟ **(added height)**
.27083333 ⊠ 95.75 ⊟ 25.932292

D) Last, calculate the partition height.
1) **(Added height)** ⊞ **(exterior-wall height)** ⊟ **(interior-wall height)**
25.932292 ⊞ 96.75 ⊟ 122.68229
2) Convert the height to fractional inches: 122.68229" = 122¹¹⁄₁₆".

The wall height is 122 ¹¹⁄₁₆".

Problems

1) You are framing an interior 2 × 8 bearing wall. The distance from the exterior of the building to the interior bearing wall is 93¾". The exterior bearing wall is a 2 × 4 wall 97" high. The roof pitch is 5/12 and the rafters are to have full bearing on both walls. What is the height of the interior bearing wall?

2) You are working on a building with a 10½/12 roof pitch. The exterior bearing wall is a 2 × 6 wall, and there is to be an interior 4 × 10 bearing beam 9'8½" from the exterior. The ridge support purlin is a 5⅛" × 15" glulam beam 17'3" from its outside face to the exterior. If the exterior wall is 12'2" high and the rafters get full bearing, what are the heights of the beams?

Worksheet

BEARING-WALL HEIGHT

A) Find the support run. (Remember to change the distance to decimal inches.)
(Distance to interior wall) ⊟ **(rafter bearing at exterior wall)** ⊞ **(rafter bearing at interior wall)** ⊟ **(support run)**
_____ ⊟ _____ ⊞ _____ ⊟ _____

B) Find the height multiplier.
(Unit rise) ⊟ **(unit run)** ⊟ **(height multiplier)**
_____ ⊟ _____ ⊟ _____

C) Find the added height.
(Height multiplier) ⊠ **(support run)** ⊟ **(added height)**
_____ ⊠ _____ ⊟ _____

D) Find the interior-wall height.
1) **(Added height)** ⊞ **(exterior-wall height)** ⊟ **(interior-wall height)**
_____ ⊞ _____ ⊟ _____
2) Convert to fractional inches.

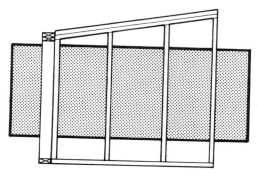

8
Rake-Wall Layout

A **rake wall** is a wall with sloping top plates and studs of different heights (Figure 8-1).

There probably is not a carpenter alive who hasn't had trouble laying out rake walls! Many carpenters avoid doing layout because they are afraid of making the rake walls too high, too low, or of the wrong pitch.

This chapter gives you an accurate, fast method of laying out any rake walls you are likely to encounter. You will find the angle to set the skill saw to cut the top of the studs, the length of the top plate, the length of each stud, and the layout, or placement, of each stud on the top plate.

Finding the layout of the top plate for each stud and the length of each stud requires repeating the same calculations over and over. Not only is it a slow process, but there is a good chance that you will push the wrong buttons by mistake. To speed up the calculations and reduce the number of opportunities for error, you can program a

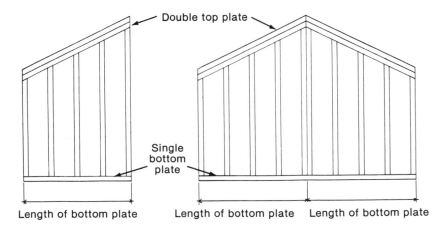

Rake walls
Figure 8-1

calculator to perform all of the repetitious calculations. Therefore, in order to perform the calculations in this chapter, you should have a programmable calculator. The calculations can be made on a non-programmable calculator, but it is time-consuming. With a programmable calculator, you will get answers as fast you can write them down. In addition, your calculator should have six memories. Again, you can manage these calculations without any memories, but you will be able to work much faster and with less confusion if your calculator does have them.

TERMS

To clarify this discussion, you need to know several terms.

Top-plate Layout. The **top-plate layout** is the distance from the end of the top plate to the closest side of the stud you are laying out (Figure 8-2). The top-plate layout on a rake wall is different from the bottom-plate layout because the top plate slopes, while the studs must remain parallel. If you get the top-plate layout wrong, you have probably made the most common mistake in building rake walls.

Stud Length. The **stud length** is the length of the stud to the short point. The short-point length is the distance from the square cut end of the stud to the closest side of the angled top cut (Figure 8-2). All the studs in a rake wall are cut on an angle to fit the sloping top plate. Since it is easier to cut from the short-point side of a stud with a skill

Rake-Wall Layout **119**

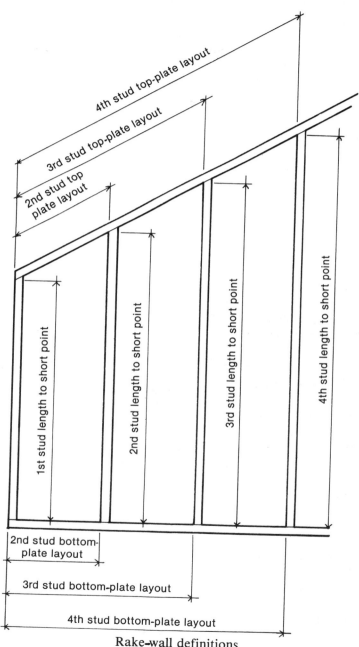

Rake-wall definitions
Figure 8-2

saw, all the stud lengths are calculated to the short point.

First-stud Length. The **first-stud length** is the length to the short point of the stud at the beginning of the rake wall (Figure 8-2). This is an important dimension because the lengths of all the other studs in the rake wall are proportional to the length of the first stud.

Memory. Memory refers to the memory of a calculator. Memory 1 is the first memory used, memory 2 is the second memory used, memory 3 is the third memory used, and so on. There are different buttons for storing numbers in memories and for recalling them. The key used on most calculators for storing a number in a memory is STO. To store a number in memory 1, use STO 1; to store a number in memory 2, use STO 2, and so on. For recalling a number from a memory, the key used on most calculators is RCL. To recall a number from memory 4, use RCL 4, to recall a number from memory 5, use RCL 5, and so on.

You push the INV tan (inverse tangent) keys to find the angle of any roof pitch given its tangent. This is the angle you will set on the skill saw to cut the top plate and the studs.

What You Need to Know

To calculate the layout for a rake wall, you need to have the following information:

- The *unit rise* of the roof. (Recall that this is the top number given in the roof-pitch ratio. If the roof pitch is $4/12$, sometimes written 4:12, the unit rise is 4.)
- The *unit run* of the roof. (Recall that this is the bottom number in the roof-pitch ratio. If the roof pitch is $4/12$, the unit run is 12.)
- The *height of the exterior bearing wall,* which can be found by measuring the exterior wall from the bottom of the bottom plate to the top of the top plate (Figure 8-3).
- The *amount of rafter bearing* at the exterior wall is the length of the seat cut of the rafter birdsmouth. This figure will be in the plans, or the rafter can be assumed to get full bearing (Figure 8-4).
- The *width of the bearing wall* is the width of the material used in the wall. For example, a 2 × 4 wall is 3½ " wide; a 2 × 6 wall is 5½ " wide.
- The *length of the bottom plate* is the distance from the end of the bottom plate to the ridge line. If there is no ridge line, this is the full length of the bottom plate (Figure 8-1).
- The *bottom-plate layout* shows where each stud is placed and

Rake-Wall Layout **121**

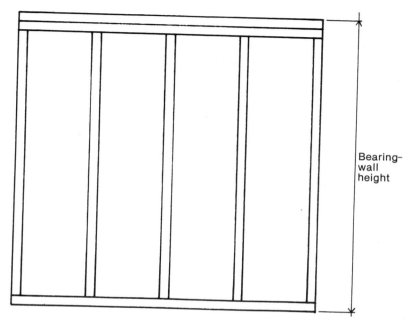

Bearing-wall height
Figure 8-3

where the ridge line is (Figure 8-5).
- The *thickness of bottom and top plates* is needed. You need to know if there are two top plates or one. If the plates are 2 × 4 or 2 × 6, they are 1½ " thick. Two top plates equal 3 ".

The length of the bottom plate and the bottom-plate layout are found by having the bottom plate cut to length and laid out. The layout should consist of a line on each side of where the stud will be placed (Figure 8-5).

You also must know how to program your calculator and how to run the program. This process varies from one brand and model of calculator to the next, so you will have to read the instruction booklet of your own calculator. It is simply a matter of pushing the correct keys and is not difficult.

Beginning Work

Once you have the eight required facts at hand, you are ready to

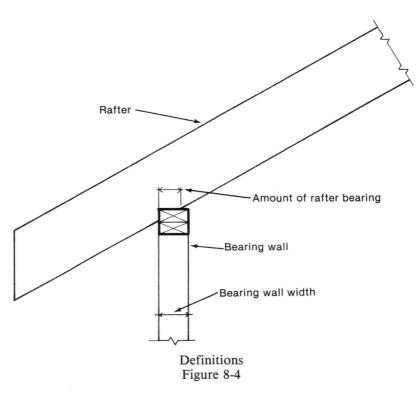

Definitions
Figure 8-4

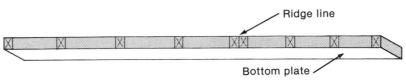

Bottom-plate layout
Figure 8-5

calculate the layout. There are seven steps, and while it will seem confusing if you just read through the steps, you will find it quite straightforward when you follow along with your own calculator. Using the program is fun, and it almost seems like magic when the answers come out! You will find that once you have gathered the necessary facts for a rake wall, it will take less than 10 minutes to calculate the angle of the cuts, the length of the top plate, the length of each stud, and the layout of each stud on the top plate.

The first step is to find the correct angle to set the skill saw in order to cut the tops of the studs correctly. To do this, divide the unit rise by the unit run and then find the inverse tangent of that number.

Next, find the **height multiplier** and the **rake multiplier** for the roof pitch on which we are working and store them in separate memories. Then, you are ready to program the calculator. You will begin with the direction, "Put the calculator in the programming mode," which means to press the buttons that prepare the calculator to learn a program. "Take the calculator out of the program mode" means to press the buttons that tell the calculator that you are finished programming it. [RUN PROGRAM] means to press the buttons that tell the calculator to perform the calculations that you have programmed.

Now you are ready to find the first-stud length. This length is found by measuring the total height of the bearing wall and storing it in a memory. Then, by subtracting the amount of rafter bearing from the width of the bearing wall (Figure 8-4) and running that figure through the program, you will tell the calculator automatically to figure the height of the rake wall and put that height in a memory. Then, to find the first-stud height, you subtract the thickness of the top and bottom plates from the height of the rake wall. To accomplish this, you must find the thickness of the top plates on a slope. Fortunately, the program will figure this for you. You need only enter the thickness of the top plate or plates and run the program. On display will be the thickness of the plates on a slope. To this, you add the thickness of the bottom plate and subtract that total from the height of the rake wall. The answer will be the first-stud height (Figure 8-6).

Finally, you are ready to find the top-plate length, the top-plate layout, and the stud heights. To do this, enter the distance from the end of the bottom plate to the ridge line on the bottom plate and run the program. On display is the top-plate length. Then, measure from the end of the bottom plate to each stud you want to figure, enter that number, and run the program. On display is the top-plate layout—the distance from the end of the top plate to the layout line for that stud (Figure 8-2). The stud length is in memory 5 and can be found by pushing [RCL] [5]. On the calculator, the process looks like this.

A) Find the angle at which to set the skill saw.
  1) **(Unit rise)** ÷ **(unit run)** [=] [INV] [tan] **(angle)**
  2) Round the angle on display to the nearest degree.

124 *Carpentry Layout*

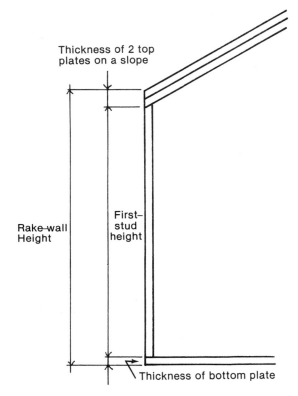

Finding the first-stud height
Figure 8-6

B) Find the height multiplier, and store it in memory 1.
   (Unit rise) ÷ **(unit run)** = **(height multiplier)** STO 1

C) Find the rake multiplier, and store it in memory 2.
   (Rise) x^2 + 12 x^2 = $\sqrt{x}$ ÷ 12 = **(rake multiplier)** STO 2

D) Program the calculator.
   1) Put the calculator in the programming mode.
   2) Enter this program:
      STO 4 × RCL 1 + RCL 3 = STO 5 RCL 4 × RCL 2 = STO 6
   3) Take the calculator out of the programming mode.

E) Now, you are ready to find the first-stud height.

1) Put the height of the bearing wall in memory 3.
   (Height of bearing wall) [STO] [3]
2) Next, find the total height of the rake wall.
   (Width of bearing wall) [−] **(amount of rafter bearing)** [=] [RUN PROGRAM] [RCL] [5] **(total height of rake wall)**
   Write down the total height of rake wall.
3) Then, find the total thickness of the top plates on a slope and the bottom plate.
   (Level thickness of top plates) [RUN PROGRAM] [+] **(thickness of bottom plate)** [=] **(total thickness of top and bottom plates)**
4) Subtract this figure from the total height of the rake wall to get the first-stud height. Store this number in memory 3.
   (Total height of rake wall) [−] **(total thickness of top and bottom plates)** [=] **(first-stud height)** [STO] [3]

F) Find the length of the top plate.
   1) **(Length of bottom plate)** [RUN PROGRAM] **(length of top plate)**
   2) Convert the length to fractional inches.

G) Find the top-plate layout and height of each stud. Repeat the following for each stud.
   1) **(Distance to stud)** [RUN PROGRAM] **(top-plate layout)**
   2) Convert the top-plate layout to fractional inches.
   3) [RCL] [5] **(stud height)**
   4) Convert the height to fractional inches.

Applications

Now you should look at some examples of this method.

Example 1

You are framing a 2 × 4 rake wall on a structure with a $\frac{2}{12}$ pitch shed roof (Figure 8-7). The bottom plate is laid out according to the diagram in Figure 8-8, and the wall will get a double top plate.
   The exterior wall is a 2 × 4 wall 96¾ " high, and the rafters will get full bearing on it. Find the top-plate length, the angle to set the skill saw to cut the top plate and studs, the top-plate layout, and the stud lengths.
   Here is what you know:
- Unit rise = 2
- Unit run = 12
- Height of the bearing wall = 96¾ "

126 *Carpentry Layout*

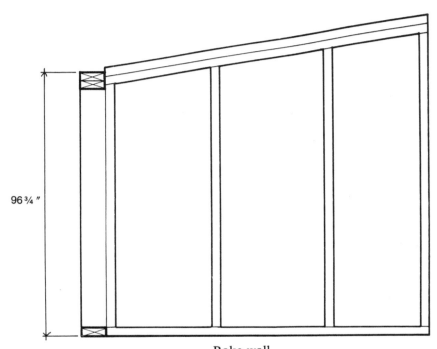

Rake wall
Figure 8-7

- Amount of rafter bearing = 3½"
- Width of the bearing wall = 3½"
- Length of the bottom plate (Figure 8-8) = 87"
- Bottom-plate layout (Figure 8-8) = 15¼", 31¼", 47¼", 63¼", 79¼", and 85½"
- Thickness of the bottom plate = 1½", and thickness of the double top plate = 3"

  Now you are ready to calculate.

A) First, find the angle at which to set the skill saw.

Rake-Wall Layout   **127**

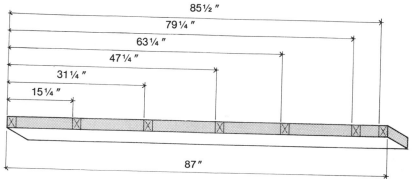

Bottom-plate layout
Figure 8-8

1) (Unit rise) ÷ **(unit run)** = INV tan **(angle)**
     2   ÷   12   = INV tan 9.4623222
2) Round to the nearest degree: 9.4623222° = 9°.

B) Find the height multiplier, and store it in memory 1.
   (Unit rise) ÷ **(unit run)** = **(height multiplier)** STO 1
      2   ÷   12   =   0.1666667   STO 1

C) Next, find the rake multiplier, and store it in memory 2.
   (Unit rise) x^2 + 12 x^2 = $\sqrt{x}$ ÷ 12 = **(rake multiplier)** STO 2
      2   x^2 + 12 x^2 = $\sqrt{x}$ ÷ 12 =   1.0137938   STO 2

D) Now, program the calculator.
   1) Put the calculator in the program mode.
   2) Enter the program.
      STO 4 × RCL 1 + RCL 3 = STO 5 RCL 4 × RCL 2 = STO 6
   3) Take the calculator out of the program mode.

E) Find the first-stud height.
   1) Store the height of the bearing wall in memory 3.
      (Height of bearing wall) STO 3
         96.75   STO 3
   2) Find the total height of the rake wall.
      (Width of bearing wall) − **(amount of rafter bearing)** = RUN PROGRAM
      RCL 5 **(total height of rake wall)**
      3.5 − 3.5 = RUN PROGRAM RCL 5 96.75

3) Find the total thickness of the top plates on a slope and the bottom plate.
   (Level thickness of top plates) [RUN PROGRAM] [+] **(thickness of bottom plate)** [=] **(total thickness of top and bottom plates)**
   3 [RUN PROGRAM] [+] 1.5 [=] 4.5413814
4) Find the first-stud height.
   (Total height of rake wall) [−] **(total thickness of top and bottom plates)** [=] **(first-stud height)** [STO] [3]
   96.75 [−] 4.5413814 [=] 92.208619 [STO] [3]

F) Now, find the length of the top plate.
   1) **(Length of bottom plate)** [RUN PROGRAM] **(length of top plate)**
      87   [RUN PROGRAM] 88.200061
   2) Convert the length to fractional inches: 88.200061" = 88 3/16."

G) Now, calculate the top-plate layout and the height of each stud. Begin with the stud at 15 1/4".
   1) **(Distance to stud)** [RUN PROGRAM] **(top-plate layout)**
      15.25   [RUN PROGRAM] 15.460355
   2) Convert the length to fractional inches: 15.460355" = 15 7/16".
   3) [RCL] [5] **(stud height)**
      [RCL] [5] 94.750285
   4) Convert the height to fractional inches: 94.750285" = 94 3/4".

   Next, figure the stud at 31 1/4".
   1) **(Distance to stud)** [RUN PROGRAM] **(top-plate layout)**
      31.25   [RUN PROGRAM] 31.681055
   2) Convert the length to fractional inches: 31.681055" = 31 11/16".
   3) [RCL] [5] **(stud height)**
      [RCL] [5] 97.416952
   4) Convert the height to fractional inches: 97.416952" = 97 7/16".

   Now, figure the stud at 47 1/4".
   1) **(Distance to stud)** [RUN PROGRAM] **(top-plate layout)**
      47.25   [RUN PROGRAM] 47.901755
   2) Convert the length to fractional inches: 47.901755" = 47 7/8".
   3) [RCL] [5] **(stud height)**
      [RCL] [5] 100.08362
   4) Convert the height to fractional inches: 100.08362" = 100 1/16".

   Next, figure the stud at 63 1/4".

1) **(Distance to stud)** [RUN PROGRAM] **(top-plate layout)**
    63.25   [RUN PROGRAM]   64.122455
2) Convert the length to fractional inches: 64.122455″ = 64⅛″.
3) [RCL] [5] **(stud height)**
    [RCL] [5]   102.75029″
4) Convert the height to fractional inches: 102.75029″ = 102¾″.

Then, figure the stud at 79¼″.
1) **(Distance to stud)** [RUN PROGRAM] **(top-plate layout)**
    79.25   [RUN PROGRAM]   80.343155
2) Convert the length to fractional inches: 80.343155″ = 80$\frac{5}{16}$″.
3) [RCL] [5] **(stud height)**
    [RCL] [5]   105.41695
4) Convert the height to fractional inches: 105.41695″ = 105$\frac{7}{16}$″.

Now, figure the last stud.
1) **(Distance to stud)** [RUN PROGRAM] **(top-plate layout)**
    85.5   [RUN PROGRAM]   86.679366
2) Convert the length to fractional inches: 86.679366″ = 86$\frac{11}{16}$″.
3) [RCL] [5] **(stud height)**
    [RCL] [5]   106.45862
4) Convert the height to fractional inches: 106.45862″ = 106$\frac{7}{16}$″.

Before you begin work, summarize your answers:
- Degree of cut = 9°
- Top-plate length = 88$\frac{3}{16}$″
- Bottom-plate layout

Bottom-plate layout	Stud height	Top-plate layout
0 (first stud)	92$\frac{3}{16}$″	0
15¼″	94¾″	15$\frac{7}{16}$″
31¼″	97$\frac{7}{16}$″	31$\frac{11}{16}$″
47¼″	100$\frac{1}{16}$″	47⅞″
63¼″	102¾″	64⅛″
79¼″	105$\frac{7}{16}$″	80$\frac{5}{16}$″
85½″	106$\frac{7}{16}$″	86$\frac{11}{16}$″

Example 2
You are framing a gable wall (Figure 8-9). The roof pitch is $\frac{10}{12}$, and the bottom plate is laid out according to Figure 8-10.

The exterior bearing walls on both sides are 2 × 6 walls, 98½″ high. The rafters bear 3½″ on each wall. All the walls have single bottom plates and double top plates. Find the top-plate lengths, the angle to set the skill saw to cut the top plate and the studs, the top-

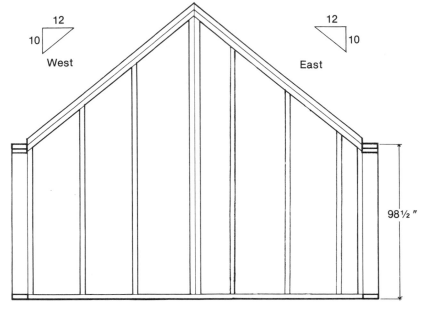

Gable-wall framing
Figure 8-9

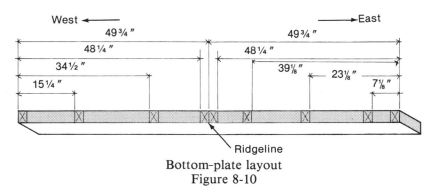

Bottom-plate layout
Figure 8-10

plate layout, and the stud lengths.
　　Here is what you know:
- Unit rise = 10
- Unit run = 12
- Height of exterior bearing wall on both sides = 98½ "
- Amount of rafter bearing on both sides = 3½ "

Rake-Wall Layout **131**

- Width of the bearing wall on both sides = 5½"
- Length of rake-wall bottom plate to the ridge line on both sides = 49¾"
- Bottom-plate layout: west = 15¼", 34½", 48¼"; east = 7⅛", 23⅛", 39⅛" 48¼"
- Thickness of the bottom plate = 1½", and double top plate = 3"

Now you can begin to calculate.

A) First find the angle at which to set the skill saw.
  1) (Unit rise) ÷ **(unit run)** = [INV] [tan] **(angle)**
     10   ÷   12   = [INV] [tan] 39.805571
  2) Round the angle to the nearest degree: 39.805571° = 40°.

B) Next, find the height multiplier, and store it in memory 1.
  (Unit rise) ÷ **(unit run)** = **(height multiplier)** [STO] [1]
     10   ÷   12   =   .8333333   [STO] [1]

C) Next, find the rake multiplier, and store it in memory 2.
  (Unit rise) [x²] [+] 12 [x²] [=] [√x] [÷] 12 [=] **(rake multiplier)** [STO] [2]
     10   [x²] [+] 12 [x²] [=] [√x] [÷] 12 [=]   1.3017083   [STO] [2]

D) Next, program the calculator.
  1) Put the calculator in the programming mode.
  2) Put the program in the calculator.
     [STO] [4] [×] [RCL] [1] [+] [RCL] [3] [=] [STO] [5] [RCL] [4] [×] [RCL] [2] [=] [STO] [6]
  3) Take the calculator out of the program mode.

E) Find the first-stud height. It will be the same on both sides for this example.
  1) Store the height of the bearing wall in memory 3.
     (Height of bearing wall) [STO] [3]
          98.5   [STO] [3]
  2) Find the total height of the rake wall.
     (Width of bearing wall) [−] **(amount of rafter bearing)** [=] [RUN PROGRAM] [RCL] [5] **(total height of rake wall)**
     5.5 [−] 3.5 [=] [RUN PROGRAM] [RCL] [5] 100.16667
  3) Find the total thickness of the top plates on a slope and the bottom plate.
     (Level thickness of top plates) [RUN PROGRAM] [+] **(thickness of bottom plate)** [=] **(total thickness of top and bottom plates)**
     3 [RUN PROGRAM] [+] 1.5 [=] 5.4051248

132 *Carpentry Layout*

    4) Find the first-stud height.
    (Total height of rake wall) ⊟ **(total thickness of top and bottom plates)** ⊜ **(first-stud height)** [STO] [3]
    100.16667 ⊟ 5.4051248 ⊜ 94.761545 [STO] [3]

F) Find the top-plate length.
    1) **(Length of bottom plate (to the ridge))** [RUN PROGRAM] **(length of top plate)**
        49.75   [RUN PROGRAM] 64.759987
    2) Convert the length to fractional inches: 64.759987″ = 64¾″.

G) Now, you can calculate the top-plate layout and the stud heights. Begin with the stud at 15¼″.
    1) **(Distance to stud)** [RUN PROGRAM] **(top-plate layout)**
        15.25   [RUN PROGRAM] 19.851051
    2) Convert the length to fractional inches: 19.851051″ = 19⅞″.
    3) [RCL] [5] **(stud height)**
        [RCL] [5] 107.46988
    4) Convert the height to fractional inches: 107.46988″ = 107½″.

Next, figure the stud at 34½″.
    1) **(Distance to stud)** [RUN PROGRAM] **(top-plate layout)**
        34.5   [RUN PROGRAM] 44.908936
    2) Convert the length to fractional inches: 44.908936″ = 44¹⁵⁄₁₆″.
    3) [RCL] [5] **(stud height)**
        [RCL] [5] 123.51155
    4) Convert the height to fractional inches: 123.51155″ = 123½″.

Now, figure the stud at 48¼″.
    1) **(Distance to stud)** [RUN PROGRAM] **(top-plate layout)**
        48.25   [RUN PROGRAM] 62.807424
    2) Convert the length to fractional inches: 62.807424″ = 62¹³⁄₁₆″.
    3) [RCL] [5] **(stud height)**
        [RCL] [5] 134.96988
    4) Convert the height to fractional inches: 134.96988″ = 135″.

In this problem, you now have to find the stud height and top-plate layout for each stud on the other side of the ridge. The first stud is the same, so you don't have to change the entry in memory 3.

    F) Start by finding the top-plate length.

1) (Length of bottom plate (to the ridge)) [RUN PROGRAM] **(length of top plate)**
               49.75   [RUN PROGRAM] 64.759987
2) Convert the length to fractional inches: 64.759987″ = 64¾″.

G) Now, you are prepared to find the top-plate layout and the height of each stud.
Start with the stud at 7⅛″.
1) (Distance to stud) [RUN PROGRAM] **(top-plate layout)**
     7.125   [RUN PROGRAM] 9.2746715
2) Convert the length to fractional inches: 9.2746715″ = 9¼″.
3) [RCL] [5] **(stud height)**
   [RCL] [5] 100.69905
4) Convert the height to fractional inches: 100.69905″ = 100¹¹⁄₁₆″.

Next, calculate the stud at 23⅛″.
1) (Distance to stud) [RUN PROGRAM] **(top-plate layout)**
    23.125   [RUN PROGRAM] 30.102004
2) Convert the length to fractional inches: 30.102004″ = 30⅛″.

3) [RCL] [5] **(stud height)**
   [RCL] [5] 114.03238
4) Convert the height to fractional inches: 114.03238″ = 114¹⁄₁₆″.

Now, figure the stud at 39⅛″.
1) (Distance to stud) [RUN PROGRAM] **(top-plate layout)**
    39.125   [RUN PROGRAM] 50.929336
2) Convert the length to fractional inches: 50.929336″ = 50¹⁵⁄₁₆″.
3) [RCL] [5] **(stud height)**
   [RCL] [5] 127.36571
4) Convert the height to fractional inches: 127.36571″ = 127⅜″.

Last, find the layout and height for the stud at 48¼″.
1) (Distance to stud) [RUN PROGRAM] **(top-plate layout)**
    48.25   [RUN PROGRAM] 62.807424
2) Convert the length to fractional inches: 62.807424″ = 62¹³⁄₁₆″.
3) [RCL] [5] **(stud height)**
   [RCL] [5] 134.96988
4) Convert the height to fractional inches: 134.96988″ = 135″.

Before you go to work, summarize your answers.

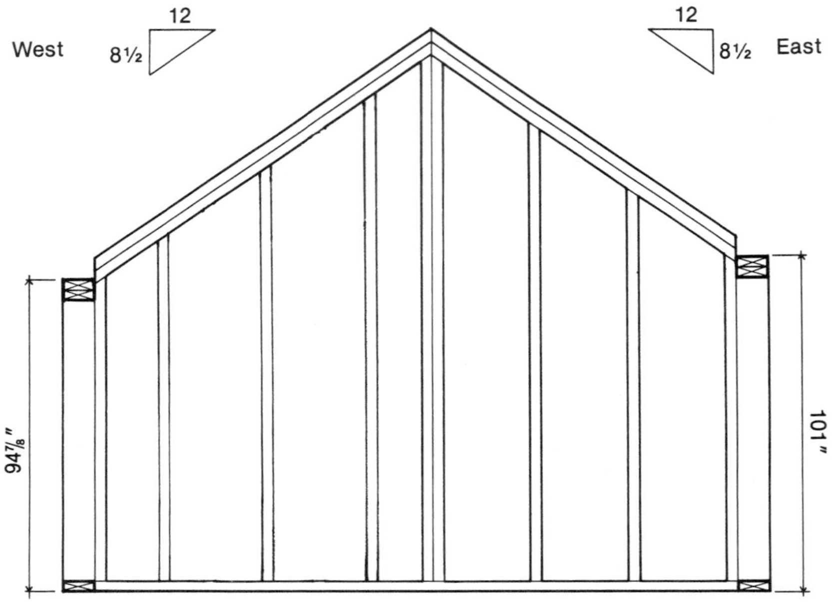

Gable wall without birdsmouths in rafters
Figure 8-11

- Degree of cut = 40°
- Top-plate length = 64¾″

• Bottom-plate layout	Stud height	Top-plate layout
West 0 (first stud)	94¾″	0
15¼″	107½″	19⅞″
34½″	123½″	44 15/16″
48¼″	135″	62 13/16″
East 0 (first stud)	94¾″	0
7⅛″	100 11/16″	9¼″
23⅛″	114 1/16″	30⅛″
39⅛″	127⅜″	50 15/16″
48¼″	135″	62 13/16″

Example 3
You are framing a gable wall (Figure 8-11). The roof pitch is 8½/12, and the bottom plate is laid out according to Figure 8-12.

The bearing wall on the east is a 2 × 6 wall, 101″ high; on the west, it is a 2 × 6 wall, 94⅞″ high. The rafters have no birdsmouths. Instead, they bear on a cant strip (Figure 8-13). All the walls get single

Rake-Wall Layout 135

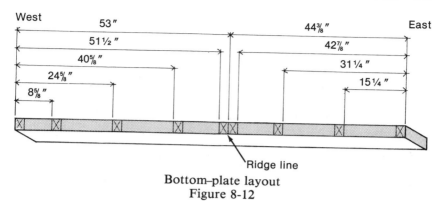

Bottom–plate layout
Figure 8-12

2 × 6 bottom plates and double 2 × 6 top plates. Find the top-plate lengths, the angle to set the skill saw to cut the top plate and studs, the top-plate layout, and the stud heights.
  Here is what you know:
- Unit rise = 8½
- Unit run = 12
- Height of exterior bearing wall: east = 101"; west = 94⅞"
- Amount of rafter bearing on both sides = 0
- Width of bearing wall on both sides = 5½"
- Length of rake-wall bottom plate: east = 44⅜"; west = 53"
- Bottom-plate layout: east = 15¼", 31¼", 42⅞"; west = 8⅝", 24⅝", 40⅝", 51½"
- Thickness of bottom plate = 1½"; double top plate = 3"

Now you can begin the calculations.

A) Find the angle to set the skill saw.
  1) **(Unit rise)** ÷ **(unit run)** = INV tan **(angle)**
      8.5   ÷   12   = INV tan   35.311213
  2) Round the angle to the nearest degree: 35.311213° = 35°.

B) Find the height multiplier, and store it in memory 1.
  (Unit rise) ÷ **(unit run)** = **(height multiplier)** STO 1
      8.5   ÷   12   =   .7083333   STO 1

C) Find the rake multiplier, and store it in memory 2.
  (Unit rise) x^2 + 12 x^2 = $\sqrt{x}$ ÷ 12 = **(rake multiplier)** STO 2
      8.5 x^2 + 12 x^2 = $\sqrt{x}$ ÷ 12 =   1.2254534   STO 2

136 *Carpentry Layout*

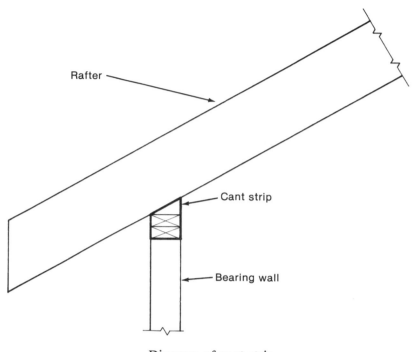

Diagram of cant strip
Figure 8-13

D) Now you can program the calculator.
 1) Put the calculator in the programming mode.
 2) Enter the program in the calculator.
    STO 4 × RCL 1 + RCL 3 = STO 5 RCL 4 × RCL 2 = STO 6
 3) Take the calculator out of the programming mode.

E) Find the first-stud height for the east rake wall.
 1) Store the height of the bearing wall in memory 3.
    (Bearing-wall height) STO 3
       101   STO 3
 2) Find the total height of the rake wall.
    (Width of bearing wall) − **(amount of rafter bearing)** = RUN PROGRAM
    RCL 5 **(total height of rake wall)**
       5.5 − 0 = RUN PROGRAM RCL 5 104.89583
 3) Find the total thickness of the top plates on a slope and the bottom plate.

(Level thickness of top plates) [RUN PROGRAM] [+] (thickness of bottom plate) [=] (total thickness of top and bottom plates)
3 [RUN PROGRAM] [+] 1.5 [=] 5.1763603

4) Find the first-stud height.
(Total height of rake wall) [−] (total thickness of top and bottom plates) [=] (first-stud height) [STO] [3]
104.89583 [−] 5.1763603 [=] 99.71947 [STO] [3]

F) Now, calculate the length of the top plate for the east rake wall.
  1) (Length of bottom plate (to the ridge)) [RUN PROGRAM] (length of top plate)
           44.375   [RUN PROGRAM] 54.379496
  2) Convert the length to fractional inches: 54.379496″ = 54⅜″.

G) Next, figure the top-plate layout and the height of each stud on the east side. Begin with the stud at 15¼″.
  1) (Distance to stud) [RUN PROGRAM] (top-plate layout)
         15.25   [RUN PROGRAM] 18.688165
  2) Convert the length to fractional inches: 18.688165″ = 18¹¹⁄₁₆″.
  3) [RCL] [5] (stud height)
     [RCL] [5] 110.52155
  4) Convert the height to fractional inches: 110.52155″ = 110½″.

Next, figure the stud at 31¼″.
  1) (Distance to stud) [RUN PROGRAM] (top-plate layout)
         31.25   [RUN PROGRAM] 38.29542
  2) Convert the length to fractional inches: 38.29542″ = 38⁵⁄₁₆″.
  3) [RCL] [5] (stud height)
     [RCL] [5] 121.85489
  4) Convert the height to fractional inches: 121.85489″ = 121⅞″.

Finally, calculate the stud at 42⅞″.
  1) (Distance to stud) [RUN PROGRAM] (top-plate layout)
         42.875   [RUN PROGRAM] 52.541316
  2) Convert the length to fractional inches: 52.541316″ = 52⁹⁄₁₆″.
  3) [RCL] [5] (stud height)
     [RCL] [5] 130.08926
  4) Convert the height to fractional inches: 130.08926″ = 130¹⁄₁₆″.

Now, turn your attention to the west side of the rake wall. Since the pitch of the roof on the east side is the same as on the west side, you don't need to refigure the height and rake multipliers.

E) You must calculate the first-stud height first.
  1) Store the height of the bearing wall in memory 3.
     (Height of bearing wall) [STO] [3]
         94.875   [STO] [3]
  2) Find the total height of the rake wall.
     (Width of bearing wall) [−] **(amount of rafter bearing)** [=] [RUN PROGRAM] [RCL] [5] **(total height of rake wall)**
     5.5 [−] 0 [=] [RUN PROGRAM] [RCL] [5] 98.770833
  3) Find the total thickness of the top plates on a slope and the bottom plate.
     (Level thickness of top plates) [RUN PROGRAM] [+] **(thickness of bottom plate)** [=] **(total thickness of plates)**
     3 [RUN PROGRAM] [+] 1.5 [=] 5.1763603
  4) Find the first-stud height.
     (Total height of rake wall) [−] **(total thickness of top and bottom plates)** [=] **(first-stud height)** [STO] [3]
     98.770833 [−] 5.1763603 [=] 93.594473 [STO] [3]

F) Next find the length of the top plate.
  1) **(Length of bottom plate (to the ridge))** [RUN PROGRAM] **(length of top plate)**
                 53   [RUN PROGRAM] 64.949032
  2) Convert the length to fractional inches: 64.949032″ = 64$\frac{15}{16}$″.

G) Find the top-plate layout and height of each stud on the west side. Begin with the stud at 8⅝″.
  1) **(Distance to stud)** [RUN PROGRAM] **(top-plate layout)**
         8.625   [RUN PROGRAM] 10.569536
  2) Convert the length to fractional inches: 10.569536″ = 10$\frac{9}{16}$″.
  3) [RCL] [5] **(stud height)**
     [RCL] [5] 99.703848
  4) Convert the height to fractional inches: 99.703848″ = 99$\frac{11}{16}$″.

  Next, figure the stud at 24⅝″.
  1) **(Distance to stud)** [RUN PROGRAM] **(top-plate layout)**
         24.625   [RUN PROGRAM] 30.176791
  2) Convert the length to fractional inches: 30.176791″ = 30$\frac{3}{16}$″.
  3) [RCL] [5] **(stud height)**
     [RCL] [5] 111.03718
  4) Convert the height to fractional inches: 111.03718″ = 111$\frac{1}{16}$″.

Now, figure the stud at 40⅝".
1) **(Distance to stud)** [RUN PROGRAM] **(top-plate layout)**
   40.625   [RUN PROGRAM]   49.784046
2) Convert the length to fractional inches: 49.784046" = 49¹³⁄₁₆".
3) [RCL] [5] **(stud height)**
   [RCL] [5]   122.37051
4) Convert the height to fractional inches: 122.37051" = 122⅜".

Finally, calculate the stud at 51½".
1) **(Distance to stud)** [RUN PROGRAM] **(top-plate layout)**
   51.5   [RUN PROGRAM]   63.110852
2) Convert the length to fractional inches: 63.110852" = 63⅛".
3) [RCL] [5] **(stud height)**
   [RCL] [5]   130.07364
4) Convert the height to fractional inches: 130.07364" = 130¹⁄₁₆".

Now summarize your answers and you're ready to work.

East side
- Degree of cut = 35°
- Length of top plate = 54⅜"
- Bottom-plate layout   Stud height   Top-plate layout
  0 (first stud)   99¹¹⁄₁₆"   0
  15¼"   110½"   18¹¹⁄₁₆"
  31¼"   121⅞"   38⁵⁄₁₆"
  42⅞"   130¹⁄₁₆"   52⁹⁄₁₆"

West side
- Degree of cut = 35°
- Length of top plate = 64¹⁵⁄₁₆"
- Bottom-plate layout   Stud height   Top-plate layout
  0 (first stud)   93⅜"   0
  8⅝"   99¹¹⁄₁₆"   10⁹⁄₁₆"
  24⅝"   111¹⁄₁₆"   30³⁄₁₆"
  40⅝"   122⅜"   49¹³⁄₁₆"
  51½"   130¹⁄₁₆"   63⅛"

Example 4
You are working on an unequal-pitch gable roof (Figure 8-14). The east side has a ⁴⁄₁₂ pitch and the west side has a ¹²⁄₁₂ pitch. The bottom plate is laid out according to Figure 8-15.

140 *Carpentry Layout*

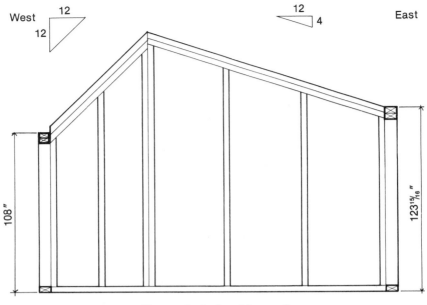

Unequal-pitch gable roof
Figure 8-14

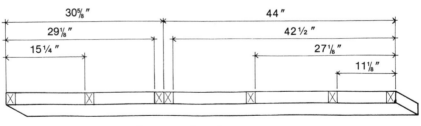

Bottom-plate layout
Figure 8-15

The exterior bearing wall on the west side is 108″ high; on the east side, it is 123$^{15}/_{16}$″ high. All the walls are 2 × 4 walls with single bottom plates and double top plates. The rafters bear fully on the east and 2″ on the west. Find all the angles and measurements needed to lay out the rake wall.

Here is what you know for the *west wall*.
- Unit rise = 12
- Unit run = 12

Rake-Wall Layout **141**

- Height of exterior bearing wall = 108″
- Amount of rafter bearing = 2″
- Width of bearing wall = 3½″
- Length of rake-wall bottom plate (to the ridge) = 30⅝″
- Bottom-plate layout = 15¼″, 29⅛″
- Thickness of bottom plate = 1½″, and of double top plate = 3″
  Now you are ready to calculate for the west side.

A) Find the angle at which to set the skill saw.
  1) (Unit rise) ÷ **(unit run)** = INV tan **(angle)**
     12   ÷   12   =   INV tan   45
  2) The angle does not need to be rounded.

B) Find the height multiplier, and store it in memory 1.
  (Unit rise) ÷ **(unit run)** = **(height multiplier)** STO 1
     12   ÷   12   =   1   STO 1

C) Find the rake multiplier, and store it in memory 2.
  (Unit rise) x^2 + 12 x^2 = $\sqrt{x}$ ÷ 12 = **(rake multiplier)** STO 2
     12   x^2 + 12 x^2 = $\sqrt{x}$ ÷ 12 =   1.4142136   STO 2

D) Now, program the calculator.
  1) Put the calculator in the programming mode.
  2) Enter the program in the calculator.
     STO 4 × RCL 1 + RCL 3 = STO 5 RCL 4 × RCL 2 = STO 6
  3) Take the calculator out of the programming mode.

E) Find the first-stud height.
  1) Store the height of the bearing wall in memory 3.
     (Height of bearing wall) STO 3
            108   STO 3
  2) Find the total height of the rake wall.
     (Width of bearing wall) − **(amount of rafter bearing)** = RUN PROGRAM
     RCL 5 **(total height of rake wall)**
     3.5 − 2 = RUN PROGRAM RCL 5 109.5
  3) Find the total thickness of the top plates on a slope and the bottom plate.
     (Level thickness of top plates) RUN PROGRAM + **(thickness of bottom plate)** = **(total thickness of top and bottom plates)**
     3 RUN PROGRAM + 1.5 = 5.7426407
  4) Find the first-stud height.

142 *Carpentry Layout*

(Total height of rake wall) $\boxminus$ (total thickness of top and bottom plates) $\boxminus$ (first-stud height) $\boxed{\text{STO}}$ $\boxed{3}$
109.5 $\boxminus$ 5.7426407 $\boxminus$ 103.75736 $\boxed{\text{STO}}$ $\boxed{3}$

F) Find the length of the top plate.
  1) (Length of bottom plate (to the ridge)) $\boxed{\text{RUN PROGRAM}}$ (length of top plate)
    30.625 $\boxed{\text{RUN PROGRAM}}$ 43.31029
  2) Convert the length to fractional inches: 43.31029″ = 43 5/16″.

G) Now, you can calculate the top-plate layout and height of each stud on the west side. Begin with the stud at 15 1/4 ″.
  1) (Distance to stud) $\boxed{\text{RUN PROGRAM}}$ (top-plate layout)
    15.25 $\boxed{\text{RUN PROGRAM}}$ 21.566757
  2) Convert the length to fractional inches: 21.566757″ = 21 9/16″.
  3) $\boxed{\text{RCL}}$ $\boxed{5}$ (stud height)
    $\boxed{\text{RCL}}$ $\boxed{5}$ 119.00736
  4) Convert the height to fractional inches: 119.00736″ = 119″.

Next, figure the stud at 29 1/8 ″.
  1) (Distance to stud) $\boxed{\text{RUN PROGRAM}}$ (top-plate layout)
    29.125 $\boxed{\text{RUN PROGRAM}}$ 41.18897
  2) Convert the length to fractional inches: 41.18897″ = 41 3/16″.
  3) $\boxed{\text{RCL}}$ $\boxed{5}$ (stud height)
    $\boxed{\text{RCL}}$ $\boxed{5}$ 132.88236
  4) Convert the height to fractional inches: 132.88236″ = 132 7/8″.

Now, you can find the layout angles and dimensions for the east side of the rake wall. Here is what you know:
- Unit rise = 4
- Unit run = 12
- Height of exterior bearing wall = 123 15/16 ″
- Amount of rafter bearing = 3 1/2 ″
- Width of bearing wall = 3 1/2 ″
- Length of rake-wall bottom plate (to the ridge) = 44 ″
- Bottom-plate layout = 11 1/8 ″, 27 1/8 ″, 42 1/2 ″
- Bottom plate = 1 1/2 ″, double top plate = 3 ″

You are ready to begin calculating.

A) Find the angle at which to set the skill saw.
  1) (Unit rise) $\boxed{\div}$ (unit run) $\boxminus$ $\boxed{\text{INV}}$ $\boxed{\text{tan}}$ (angle)
    4    $\boxed{\div}$ 12 $\boxminus$ $\boxed{\text{INV}}$ $\boxed{\text{tan}}$ 18.434949

Rake-Wall Layout

2) Round the angle to the nearest degree: 18.434949° = 18°.

B) Find the height multiplier, and store it in memory 1.
(**Unit rise**) ÷ 12 = (**height multiplier**) STO 1
4 ÷ 12 = .3333333 STO 1

C) Find the rake multiplier, and store it in memory 2.
(**Unit rise**) x^2 + (**unit run**) x^2 = $\sqrt{x}$ ÷ 12 = (**rake multiplier**) STO 2
4 x^2 + 12 x^2 = $\sqrt{x}$ ÷ 12 = 1.0540926 STO 2

D) Program the calculator.
1) Put the calculator in the programming mode.
2) Enter the program.
STO 4 × RCL 1 + RCL 3 = STO 5 RCL 4 × RCL 2 = STO 6
3) Take the calculator out of the programming mode.

E) Find the first-stud height.
1) Store the bearing-wall height in memory 3.
(**Height of bearing wall**) STO 3
123.9375 STO 3
2) Find the total height of the rake wall.
(**Width of bearing wall**) − (**amount of rafter bearing**) = RUN PROGRAM
RCL 5 (**total height of rake wall**)
3.5 − 3.5 = RUN PROGRAM RCL 5 123.9375
3) Find the total thickness of the top plates on a slope and the bottom plate.
(**Level thickness of top plates**) RUN PROGRAM + (**thickness of bottom plate**) = (**total thickness of top and bottom plates**)
3 RUN PROGRAM + 1.5 = 4.6622777
4) Find the first-stud height.
(**Total height of rake wall**) − (**total thickness of top and bottom plates**) = (**first-stud height**) STO 3
123.9375 − 4.6622777 = 119.27522 STO 3

F) Find the length of the top plate.
1) (**Length of bottom plate**) RUN PROGRAM (**length of top plate**)
44 RUN PROGRAM 46.380072
2) Convert the length to fractional inches: 46.380072″ = 46⅜″.

G) Now, calculate the top-plate layout and the height of each stud on the east side. Begin with the stud at 11⅛″.
1) (**Distance to stud**) RUN PROGRAM (**top-plate layout**)
11.125 RUN PROGRAM 11.72678

2) Convert the length to fractional inches: 11.72678″ = 11¾″.
3) [RCL] [5] (stud height)
   [RCL] [5] 122.98356
4) Convert the height to fractional inches: 122.98356″ = 123″.

Now, calculate the stud at 27⅛″.
1) **(Distance to stud)** [RUN PROGRAM] **(top-plate layout)**
   27.125   [RUN PROGRAM] 28.592261
2) Convert the length to fractional inches: 28.592261″ = 28⁹⁄₁₆″.
3) [RCL] [5] **(stud height)**
   [RCL] [5] 128.31689
4) Convert the height to fractional inches: 128.31689″ = 128⁵⁄₁₆″.

Next, calculate the stud at 42½″.
1) **(Distance to stud)** [RUN PROGRAM] **(top-plate layout)**
   42.5   [RUN PROGRAM] 44.798934
2) Convert the length to fractional inches: 44.798934″ = 44¹³⁄₁₆″.
3) [RCL] [5] **(stud height)**
   [RCL] [5] 133.44189
4) Convert the height to fractional inches: 133.44189″ = 133⁷⁄₁₆″.

Now, summarize your answers.

West side
- Degree of cut = 45°
- Length of top plate = 43⁵⁄₁₆″

Bottom-plate layout	Stud height	Top-plate layout
0 (first stud)	103¾″	0
15¼″	119″	21⁹⁄₁₆″
29⅛″	132⅞″	41³⁄₁₆″

East side
- Degree of cut = 18°
- Length of top plate = 46⅜″

Bottom-plate layout	Stud height	Top-plate layout
0 (first stud)	119¼″	0
11⅛″	123″	11¾″
27⅛″	128⁵⁄₁₆″	28⁹⁄₁₆″
42½″	133⁷⁄₁₆″	44¹³⁄₁₆″

Rake-Wall Layout 145

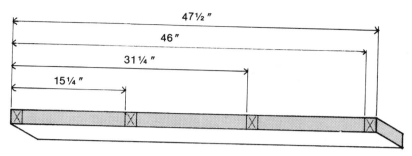

Bottom–plate layout
Figure 8-16

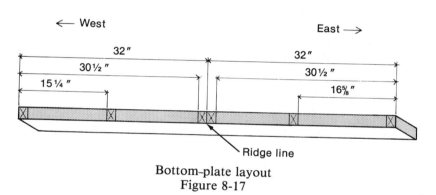

Bottom–plate layout
Figure 8-17

Problems

1) You are framing a 2 × 4 rake wall on a structure with a $3/_{12}$ pitch shed roof. The bottom plate is laid out according to Figure 8-16 and will get a double top plate. The exterior wall is 96" high, and the rafters will get full bearing on it. Find all the information you need to lay out and cut the rake wall.

2) You are framing a gable wall with the bottom plate laid out according to Figure 8-17. The roof pitch is $9¼/12$ and the 2 × 6 exterior bearing walls on each side are 10' high. The rafters bear 3" on each side, and the walls have single bottom plates and double top plates. Find all the information you need to lay out and cut the gable wall.

Worksheet

RAKE-WALL LAYOUT

A) Find the angle at which to set the skill saw.
1) **(Unit rise)** $\div$ **(unit run)** $=$ **INV** **tan** **(angle)**
   _____ $\div$ _____ $=$ **INV** **tan** _____
2) Round the angle in display to the nearest degree.

B) Find the height multiplier, and store it in memory 1.
(Unit rise) $\div$ **(unit run)** $=$ **(height multiplier)** **STO** **1**
_____ $\div$ _____ $=$ _____ **STO** **1**

C) Find the rake multiplier, and store it in memory 2.
(Rise) x^2 $+$ 12 x^2 $=$ $\sqrt{x}$ $\div$ 12 $=$ **(rake multiplier)** **STO** **2**
_____ x^2 $+$ 12 x^2 $=$ $\sqrt{x}$ $\div$ 12 $=$ _____ **STO** **2**

D) Program the calculator.
1) Put the calculator in the programming mode.
2) Enter this program:
   STO **4** $\times$ **RCL** **1** $+$ **RCL** **3** $=$ **STO** **5** **RCL** **4** $\times$ **RCL** **2** $=$ **STO** **6**
3) Take the calculator out of the programming mode.

E) Find the first-stud height.
1) Store the height of the bearing wall in memory 3.
   (Height of bearing wall) **STO** **3**
   _____ **STO** **3**
2) Find the total height of the rake wall.
   (Width of bearing wall) $-$ **(amount of rafter bearing)** $=$ **RUN PROGRAM** **RCL** **5** **(total height of rake wall)**
   _____ $-$ _____ $=$ **RUN PROGRAM** **RCL** **5** _____
3) Find the total thickness of the top plates on a slope and the bottom plate.
   (Level thickness of top plates) **RUN PROGRAM** $+$ **(thickness of bottom plate)** $=$ **(total thickness of top and bottom plates)**
   _____ **RUN PROGRAM** $+$ _____ $=$ _____
4) Find the first-stud height.
   (Total height of rake wall) $-$ **(total thickness of top and bottom plates)** $=$ **(first-stud height)** **STO** **3**
   _____ $-$ _____ $=$ _____ **STO** **3**

F) Now, find the length of the top plate.

1) **(Length of bottom plate)** [RUN PROGRAM] **(length of top plate)**
   _____ [RUN PROGRAM] _____
2) Convert the length to fractional inches.

G) Find the top-plate layout and height of each stud. Repeat the following for each stud.
   1) **(Distance to stud)** [RUN PROGRAM] **(top-plate layout)**
      _____ [RUN PROGRAM] _____
   2) Convert the length to fractional inches.
   3) [RCL] [5] **(stud height)**
      [RCL] [5] _____
   4) Convert the height to fractional inches.

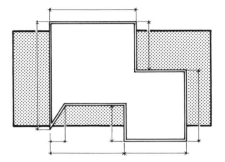

9
Calculating Foundation Layouts

Many carpenters find it difficult to lay out foundations, especially if the layouts are complicated. Sometimes there are so many batterboards and strings that confusion is the only result. On very steep hills, the batterboards can be so high that you need a ladder to reach them.

Tools You'll Need

In this chapter, you will learn a method of laying out the foundation corner stakes without using batterboards. All that you will need is an accurate transit, a long tape measure in decimal feet (feet and tenths instead of feet and inches), and a foundation plan. If the transit can be read to the nearest 30 seconds, you can be accurate to $\frac{3}{16}$" in 100'. This is all the accuracy you will need to lay out foundations.

All the corners on a foundation plan can be described in terms of an angle and a distance from one point. This means that you will be

Calculating Foundation Layouts **149**

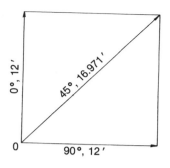 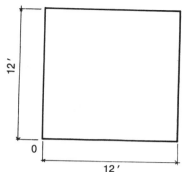

Methods of locating foundation corners
Figure 9-1

able to set the transit up over one point (the control point), turn the transit to the calculated angle, and measure the distance to any of the other points you need to find (Figure 9-1). The calculations in this chapter convert the dimensions on the foundation plan to the angles and distances needed to lay out a foundation using this method.

Using Your Calculator

To perform these calculations, your calculator must have the following keys:

- [R▸P] This is is the polar coordinate function. It converts the dimensions on the foundation plan to an angle and a distance. On some calculators this key is shown as [P▸R]. In this case, you will need to use the [INV] (inverse) key with it. The calculator I used for this book has a [P▸R] key, so I will be using the [INV] key.
- [DD-DMS] This is the function that converts decimal degrees to degrees minutes, and seconds. Some calculators—for example, the one I use—has a [DMS-DD] key and you will need to use the [INV] key as I do. Most calculators work in decimal degrees, but transits are usually calibrated in degrees, minutes, and seconds. This function allows you to convert from the decimal degrees on the calculator to the degrees, minutes, and seconds you will need for the transit. Some calculators can calculate in degrees, minutes, and seconds; with these calculators you will not need to convert the answer.
- [x≠y] This is the X exchange Y key. It allows you to enter the dimensions on the foundation plan. On some calculators this key is ENTER or ENTER X.

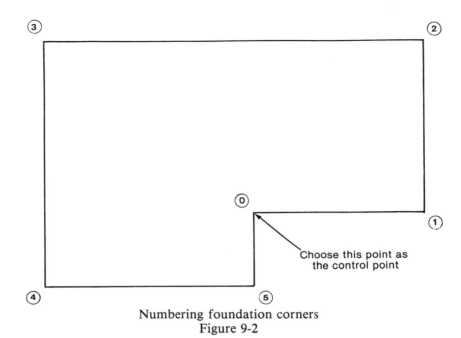

Numbering foundation corners
Figure 9-2

- ⊞ This is a key that changes signs. By entering a number and pressing this key, you can enter a negative number.

Beginning Work

With these functions, you can calculate the angles and distances you'll need to lay out the foundation. Begin by converting all the dimensions on the foundation plan from feet, inches, and fractions of inches to decimal feet. To accomplish this, divide the numerator (top number) of the fraction by the denominator (bottom number). Then add the inches and divide by 12. Add the feet, and you have converted to decimal feet.

Now, number all the corners on the foundation plan that you want to lay out. Use point 0 as the control point; this is the corner on which you will set the transit, so choose a corner with an unobstructed view of all the other corners. If the building is long, try to choose a point that is central in order to avoid long pulls with the tape measure (Figure 9-2).

Next, set up a point chart. On it, put all the distances (in decimal feet) on the foundation plan to each point from point 0. The chart should have five columns (Figure 9-3). The first column (*Corner*) lists

Corner	Distance up or down	Distance right or left	Angle	Distance from Point 0
1				
2				
3				
4				
5				
6				
7				
8				
9				
10				
•				
•				
•				

Point chart
Figure 9-3

the corners. The second column (*Distance up or down*) is the distance up or down from point 0 to each of the corners listed. If the corner is upward on the drawing from point 0, the distance is a positive number. If it is downward from point 0, the distance is a negative number. The third column (*Distance right or left*) is the distance to the right or left from point 0 of each corner. If the corner is to the left of point 0 on the drawing, the distance is a negative number. If the corner is to the right, the distance is a positive number. All the entries in the distance columns are in decimal feet. The fourth column *(Angle)* lists the angle to each corner from point 0 in degrees, minutes, and seconds. The fifth column (*Distance from point 0*) is the distance to each corner from point 0 in decimal feet rounded to three places. You don't need to convert it to feet and inches because most long tapes are calibrated in decimal feet.

You must fill in the angle and distance columns for each layout. To do this, set the angle for a point by entering the number out of the "distance up or down" column and pressing the [x≠y] key. You then enter the number from the "distance right or left" column and press the [INV] [P▸R] key. This gives the angle in decimal degrees. Press the [INV] [DMS-DD] key to display the angle in degrees, minutes, and seconds. Write this in the angle column. Press the [x⇌y] key to display the distance in decimal feet. Round this to three places and write it in the distance column.

On the calculator, here is the sequence you will follow.

A) Convert the feet, inches, and fractions on the foundation plan to decimal feet.
 1) **(Numerator)** [÷] **(denominator)** [+] **(in.)** [=] [÷] 12 [+] **(ft.)** [=] **(decimal ft.)**
 2) Round the number to three decimal places.

B) Set up a point chart and fill in the numbers you know.

C) Calculate the angle and distance from point 0 to each of the other corners by repeating the following sequence for each corner.
 1) Calculate the angle.
 (Distance up or down) [x⇌y] **(distance right or left)** [INV] [P▸R] **(angle in decimal degrees)**
 2) Convert from decimal degrees to degrees, minutes, and seconds.
 (Decimal degrees) [INV] [DMS-DD] **(degrees, minutes, and seconds)**
 3) Find the distance from point 0 to the point.
 [x⇌y] **(distance in decimal feet)**
 4) Round the distance to three decimal places.

Applications

Now, look at some examples, to illustrate this method.

Example 1

You are working on a foundation with the plan shown in Figure 9-4. Find all the dimensions and angles necessary to complete a point chart.

First, label every corner on the foundation plan using point 0 as the control point (Figure 9-5).

A) The first computation step is to convert all the dimensions in feet and inches on the foundation plan to decimal feet. In this example, none of the dimensions need to be converted.

B) Set up the point chart and fill in the spaces that you know.

Calculating Foundation Layouts 153

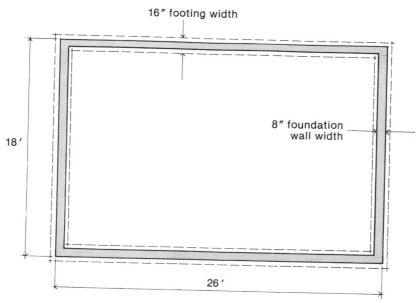

Foundation plan
Figure 9-4

Corner	Distance up or down	Distance right or left	Angle	Distance from 0
1	0	26		
2	18	26		
3	18	0		

C) Next, calculate the angle and distance from point 0 to each of the other corners.
Begin by finding the angle and distance to point 1.
  1) Calculate the angle.
   (Distance up or down) [x≠y] **(distance right or left)** [INV] [P→R] **(angle in decimal degrees)**
   0 [x≠y] 26 [INV] [P→R] 90
  2) Convert the decimal angle to degrees, minutes, and seconds.
   (Decimal degrees) [INV] [DMS-DD] **(degrees, minutes, seconds)**
       90   [INV] [DMS-DD] 90
  3) Find the distance to point 1.
   [x≠y] **(distance)**
   [x≠y] 26

154 *Carpentry Layout*

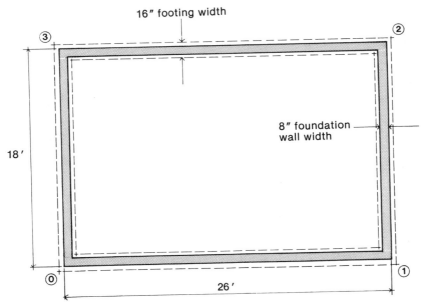

Foundation plan with numbered corners
Figure 9-5

Next, calculate the angle and distance from point 0 to point 2.
1) Calculate the angle.
   (Distance up or down) [x≠y] **(distance right or left)** [INV] [P▸R] **(angle in decimal degrees)**
   18 [x≠y] 26 [INV] [P▸R] 55.304846
2) Convert to degrees, minutes, and seconds.
   (Decimal degrees) [INV] [DMS-DD] **(degrees, minutes, seconds)**
   55.304846   [INV] [DMS-DD] 55.181745
   This is equivalent to 55°18′17″.
3) Find the distance.
   [x≠y] **(distance)**
   [x≠y] 31.622777
4) Round off to three places: 31.622777′ = 31.623′.

Calculate the angle and distance to point 3.
1) Calculate the angle.
   (Distance up or down) [x≠y] **(distance right or left)** [INV] [P▸R] **(angle in decimal degrees)**   18 [x≠y] 0 [INV] [P▸R] 0

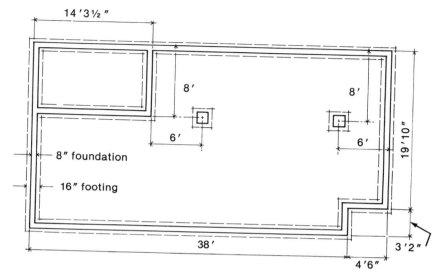

Foundation plan
Figure 9-6

2) Convert to degrees, minutes, and seconds.
   (Decimal degrees) INV DMS-DD **(degrees, minutes, seconds)**
   0   INV DMS-DD 0
3) Find the distance from point 0 to point 3.
   x≠y **(distance)**
   x≠y 18

Write these numbers in the point chart.

Corner	Distance up or down	Distance right or left	Angle	Distance from point 0
1	0'	26'	90°	26'
2	18'	26'	55°18'17"	31.628'
3	18'	0'	0°	18'

All the points are now calculated. Set the transit up over the control point and shoot in the foundation.

Example 2
Using a foundation plan as in Figure 9-6, find all the dimensions and angles necessary to shoot in every corner and the interior piers with a transit.

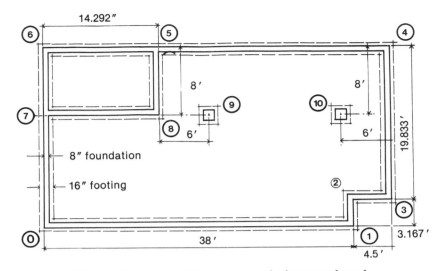

Foundation plan with corners and piers numbered
Figure 9-7

First, label every corner with a number, using point 0 as the control point (Figure 9-7).

A) Convert the feet, inches, and fractions on the foundation plan to decimal feet.

Convert 3'2" to decimal feet.
1) **(In.)** ÷ 12 + **(ft.)** = **(decimal ft.)**
   2 ÷ 12 + 3 = 3.1666667
2) Round to three decimal places: 3.1666667' = 3.167'.

Convert 4'6" to decimal feet.
1) **(In.)** ÷ 12 + **(ft.)** = **(decimal ft.)**
   6 ÷ 12 + 4 = 4.5

Convert 19'10" to decimal feet.
1) **(In.)** ÷ 12 + **(ft.)** = **(decimal ft.)**
   10 ÷ 12 + 19 = 19.833333
2) Round to three decimal places: 19.833333' = 19.833'.

Convert 14'3½" to decimal feet.
1) **(Numerator)** ÷ **(denominator)** + **(in.)** = ÷ 12 + **(ft.)** = **(decimal ft.)**
   1 ÷ 2 + 3 = ÷ 12 + 14 = 14.291667
2) Round the number to three decimal places: 14.291667' = 14.292'.

Put the dimensions in decimal feet on your drawing, as in Figure 9-7.

B) Next, make a point chart and fill in the numbers you know.

Corner	Distance up or down	Distance right or left	Angle	Distance from point 0
1	0	38		
2	3.167	38		
3	3.167	42.5		
4	23	42.5		
5	23	14.292		
6	23	0		
7	15	0		
8	15	14.292		
9	15	20.292		
10	15	36.5		

C) Now, you are ready to calculate the angles and distances.
First, find the angle and distance from point 0 to point 1:
1) Calculate the angle.
   (Distance up or down) [x⇔y] **(distance right or left)** [INV] [P▶R] **(angle in decimal degrees)**
   0 [x⇔y] 38 [INV] [P▶R] 90
2) Convert to degrees, minutes, and seconds.
   (Decimal degrees) [INV] [DMS-DD] **(degrees, minutes, seconds)**
   90   [INV] [DMS-DD] 90
   This is 90°.
3) Find the distance.
   [x⇔y] **(distance)**
   [x⇔y] 38

Now, calculate the angle and distance from point 0 to point 2.
1) Calculate the angle.
   (Distance up or down) [x⇔y] **(distance right or left)** [INV] [P▶R] **(angle in decimal degrees)**
   3.167 [x⇔y] 38 [INV] [P▶R] 85.235859
2) Convert to degrees, minutes, and seconds.
   (Decimal degrees) [INV] [DMS-DD] **(degrees, minutes, seconds)**
   85.235859   [INV] [DMS-DD] 85.140909
   This is the angle 85°14′9″.
3) Find the distance.

158 *Carpentry Layout*

      x⇔y (distance)
      x⇔y 38.131744
   4) Round the distance to three decimal places: 38.131744' = 38.132'.

Calculate the angle and distance from point 0 to point 3.
   1) Calculate the angle.
      (Distance up or down) x⇔y **(distance right or left)** INV P▸R **(angle in decimal degrees)**
      3.167 x⇔y 42.5 INV P▸R 85.73833
   2) Convert to degrees, minutes, and seconds.
      (Decimal degrees) INV DMS-DD **(degrees, minutes, seconds)**
      85.73833   INV DMS-DD 85.441799
      This is the angle 85°44'18".
   3) Find the distance.
      x⇔y (distance)
      x⇔y 42.617835
   4) Round the distance to three decimal places: 42.617835' = 42.618'.

Calculate the angle and distance from point 0 to point 4.
   1) Calculate the angle.
      (Distance up or down) x⇔y **(distance right or left)** INV P▸R **(angle in decimal degrees)**
      23 x⇔y 42.5 INV P▸R 61.578791
   2) Convert to degrees, minutes, and seconds.
      (Decimal degrees) INV DMS-DD **(degrees, minutes, seconds)**
      61.578791   INV DMS-DD 61.344365
      This is the angle 61°34'44".
   3) Find the distance.
      x⇔y **(distance)**
      x⇔y 48.324424
   4) Round the distance to three decimal places: 48.324424' = 48.324'.

Calculate the angle and distance from point 0 to point 5.
   1) Calculate the angle.
      (Distance up or down) x⇔y **(distance right or left)** INV P▸R **(angle in decimal degrees)**
      23 x⇔y 14.292 INV P▸R 31.856458
   2) Convert to degrees, minutes, and seconds.
      (Decimal degrees) INV DMS-DD **(degrees, minutes, seconds)**
      31.856458   INV DMS-DD 31.512325
      This angle is 31°51'23".
   3) Find the distance.

[x≠y] **(distance)**
[x≠y] 27.078797
4) Round the distance to three decimal places: 27.078797′ = 27.079′.

Calculate the angle and distance from point 0 to point 6.
1) Calculate the angle.
(Distance up or down) [x≠y] **(distance right or left)** [INV] [P→R] **(angle in decimal degrees)**
23 [x≠y] 0 [INV] [P→R]   0
2) Convert to degrees, minutes, and seconds.
(Decimal degrees) [INV] [DMS-DD] **(degrees, minutes, seconds)**
     0   [INV] [DMS-DD]   0
This angle is 0°.
3) Find the distance.
[x≠y] **(distance)**
[x≠y]   23

Calculate the angle and distance from point 0 to point 7.
1) Calculate the angle.
(Distance up or down) [x≠y] **(distance right or left)** [INV] [P→R] **(angle in decimal degrees)**
15 [x≠y] 0 [INV] [P→R]   0
2) Convert to degrees, minutes, and seconds.
(Decimal degrees) [INV] [DMS-DD] **(degrees, minutes, seconds)**
     0   [INV] [DMS-DD]   0
This angle is 0°.
3) Find the distance.
[x≠y] **(distance)**
[x≠y]   15

Find the angle and distance from point 0 to point 8.
1) Calculate the angle.
(Distance up or down) [x≠y] **(distance right or left)** [INV] [P→R] **(angle in decimal degrees)**
15 [x≠y] 14.292 [INV] [P→R]   43.615406
2) Convert to degrees, minutes, and seconds.
(Decimal degrees) [INV] [DMS-DD] **(degrees, minutes, seconds)**
   43.615406   [INV] [DMS-DD]   43.365546
This angle is 43°36′55″.
3) Find the distance.
[x≠y] **(distance)**
[x≠y]   20.718621

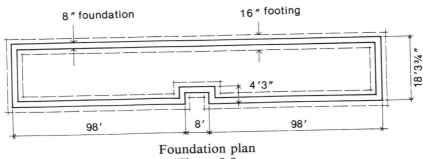

Foundation plan
Figure 9-8

    4) Round the distance to three decimal places: 20.718621' = 20.719'.

Find the angle and distance from point 0 to point 9.
    1) Calculate the angle.
        (Distance up or down) [x⇔y] **(distance right or left)** [INV] [P►R] **(angle in decimal degrees)**
        15 [x⇔y] 20.292 [INV] [P►R] 53.527908
    2) Convert to degrees, minutes, and seconds.
        (Decimal degrees) [INV] [DMS-DD] **(degrees, minutes, seconds)**
        53.527908   [INV] [DMS-DD]   53.314047
        This angle is 53°31'40".
    3) Find the distance.
        [x⇔y] **(distance)**
        [x⇔y] 25.234208
    4) Round the distance to three decimal places: 25.234208' = 25.234'.

Find the angle and distance from point 0 to point 10.
    1) Calculate the angle.
        (Distance up or down) [x⇔y] **(distance right or left)** [INV] [P►R] **(angle in decimal degrees)**
        15 [x⇔y] 36.5 [INV] [P►R] 67.659352
    2) Convert to degrees, minutes, and seconds.
        (Decimal degrees) [INV] [DMS-DD] **(degrees, minutes, seconds)**
        67.659352   [INV] [DMS-DD]   67.393367
        This angle is 67°39'34".
    3) Find the distance.
        [x⇔y] **(distance)**
        [x⇔y] 39.462007
    4) Round the distance to three decimal places: 39.462007' = 39.462'.

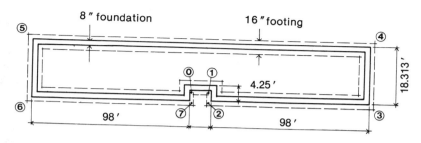

Foundation plan in decimal feet with corners numbered
Figure 9-9

Now, you can complete the point chart—it looks like this.

Corner	Distance up or down	Distance right or left	Angle	Distance from point 0
1	0'	38'	90°	38'
2	3.167'	38'	85°14'9"	38.132'
3	3.167'	42.5'	85°44'18"	42.618'
4	23'	42.5'	61°34'44"	48.324'
5	23'	14.292'	31°51'23"	27.079'
6	23'	0'	0°	23'
7	15'	0'	0°	15'
8	15'	14.292'	43°36'55"	20.719'
9	15'	20.292'	53°31'40"	25.234'
10	15'	36.5'	67°39'34"	39.462'

You are ready to set the transit up over point 0 and shoot in the foundation.

Example 3

You are required to lay out a foundation from the plan in Figure 9-8.
Find all the angles and distances necessary to lay out the foundation with a transit.
Begin by labeling every corner with a number (Figure 9-9).

A) Convert the dimensions to decimal feet. Put these dimensions on the drawing (Figure 9-9).
Convert 4'3" to decimal feet.
1) **(In.)** ÷ 12 + **(ft.)** = **(decimal ft.)**
   3 ÷ 12 + 4 = 4.25

Convert 18′3¾″ to decimal feet.
1) **(Numerator)** ÷ **(denominator)** + **(in.)** = ÷ 12 + **(ft.)** = **(decimal ft.)**
    3 ÷ 4 + 3 = ÷ 12 + 18 = 18.3125
2) Round the number to three decimal places: 18.3125′ = 18.313′.

B) Next, make a point chart and fill in the numbers you know.

Corner	Distance up or down	Distance right or left	Angle	Distance from point 0
1	0	8		
2	−4.25	8		
3	−4.25	106		
4	14.0625	106		
5	14.0625	−98		
6	−4.25	−98		
7	−4.25	0		

C) Now, you are ready to calculate the angles and distances.
First, find the angle and distance from point 0 to point 1:
1) Calculate the angle.
(Distance up or down) [x⇌y] **(distance right or left)** [INV] [P→R] **(angle in decimal degrees)**
0 [x⇌y] 8 [INV] [P→R] 90
2) Convert to degrees, minutes, and seconds.
(Decimal degrees) [INV] [DMS-DD] **(degrees, minutes, seconds)**
    90   [INV] [DMS-DD]   90
The angle is 90°.
3) Find the distance.
[x⇌y] **(distance)**
[x⇌y] 8

Now, find the angle and distance from point 0 to point 2.
1) Calculate the angle.
(Distance up or down) [x⇌y] **(distance right or left)** [INV] [P→R] **(angle in decimal degrees)**
4.25 [+/−] [x⇌y] 8 [INV] [P→R] 117.97947
2) Convert to degrees, minutes, and seconds.
(Decimal degrees) [INV] [DMS-DD] **(degrees, minutes, seconds)**
    117.97947   [INV] [DMS-DD]   117.58461
This angle is 117°58′46″.
3) Find the distance.
[x⇌y] **(distance)**
[x⇌y] 9.0588355
4) Round the distance to three decimal places: 9.0588355′ = 9.059′.

Calculating Foundation Layouts **163**

Find the angle and distance from point 0 to point 3.
1) Calculate the angle.
(Distance up or down) [x⇌y] **(distance right or left)** [INV] [P▸R] **(angle in decimal degrees)**
4.25 [+/−] [x⇌y] 106 [INV] [P▸R] 92.296007
2) Convert to degrees, minutes, and seconds.
(Decimal degrees) [INV] [DMS-DD] **(degrees, minutes, seconds)**
92.296007   [INV] [DMS-DD] 92.174562
This angle is 92°17′46″.
3) Find the distance.
[x⇌y] **(distance)**
[x⇌y] 106.08517
4) Round the distance to three decimal places: 106.08517′ = 106.085′.

Find the angle and distance from point 0 to point 4.
1) Calculate the angle.
(Distance up or down) [x⇌y] **(distance right or left)** [INV] [P▸R] **(angle in decimal degrees)**
14.0625 [x⇌y] 106 [INV] [P▸R] 82.442978
2) Convert to degrees, minutes, and seconds.
(Decimal degrees) [INV] [DMS-DD] **(degrees, minutes, seconds)**
82.442978   [INV] [DMS-DD] 82.263472
The angle is 82°26′35″.
3) Find the distance.
[x⇌y] **(distance)**
[x⇌y] 106.92873
4) Round the distance to three decimal places: 106.92873′ = 106.929′.

Find the angle and distance from point 0 to point 5.
1) Calculate the angle.
(Distance up or down) [x⇌y] **(distance right or left)** [INV] [P▸R] **(angle in decimal degrees)**
14.0625 [x⇌y] 98 [+/−] [INV] [P▸R] −81.834091
Some calculators will occasionally display a negative number. If this is the case, add the number to 360: −81.834091 [+] 360 [=] 278.16591
2) Convert to decimal degrees, minutes, and seconds.
(Decimal degrees) [INV] [DMS-DD] **(degrees, minutes, seconds)**
278.16591   [INV] [DMS-DD] 278.09573
This angle is 278°9′57″.
3) Find the distance.
[x⇌y] **(distance)**
[x⇌y] 99.003808
4) Round the distance to three decimal places: 99.003808′ = 99.004′.

164 *Carpentry Layout*

Find the angle and distance from point 0 to point 6.
   1) Calculate the angle.
      (Distance up or down) [x⇌y] **(distance right or left)** [INV] [P▸R] **(angle in decimal degrees)**
      4.25 [+/−] [x⇌y] 98 [+/−] [INV] [P▸R] −92.48321
      Since this number is negative, add 360 to it:
        −92.48321 [+] 360 [=] 267.51679
   2) Convert to degrees, minutes, and seconds.
      (Decimal degrees) [INV] [DMS-DD] **(degrees, minutes, seconds)**
      267.51679   [INV] [DMS-DD]   267.31004
      The angle is 267°31′.
   3) Find the distance.
      [x⇌y] **(distance)**
      [x⇌y] 98.092112
   4) Round the distance to three decimal places: 98.092112′ = 98.092′.

Find the angle and distance from point 0 to point 7.
   1) Calculate the angle.
      (Distance up or down) [x⇌y] **(distance right or left)** [INV] [P▸R] **(angle in decimal degrees)**
      4.25 [+/−] [x⇌y] 0 [INV] [P▸R] 180°
   2) Convert to degrees, minutes, and seconds.
      (Decimal degrees) [INV] [DMS-DD] **(degrees, minutes, seconds)**
      180   [INV] [DMS-DD]   180
      The angle is 180°.
   3) Find the distance.
      [x⇌y] **(distance)**
      [x⇌y] 4.25

Now you can complete the point chart.

Corner	Distance up or down	Distance right or left	Angle	Distance from point 0
1	0	8′	90°	8′
2	−4.25′	8′	117°58′46″	9.059′
3	−4.25′	106′	92°17′46″	106.085′
4	14.0625′	106′	82°26′35″	106.929′
5	14.0625′	−98′	278° 9′57″	99.004′
6	−4.25′	−98′	267°31′	98.092′
7	−4.25′	0	180°	4.25′

You are ready to set the transit up over point 0 and shoot in the foundation.

Calculating Foundation Layouts 165

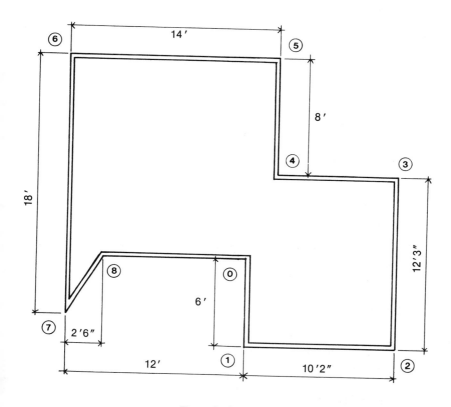

Foundation plan
Figure 9-10

Problem

You must lay out a foundation from the plan in Figure 9-10. Fill out a point chart for this foundation.

Worksheet

FOUNDATION-LAYOUT

Label every corner on the foundation plan using 0 as the control point.

A) Convert the feet, inches, and fractions on the foundation plan to decimal feet.
  1) **(Numerator)** ÷ **(denominator)** + **(in.)** = ÷ 12 + **(ft.)** = **(decimal ft.)**
     _____ ÷ _____ + ____ = ÷ 12 + ____ = _____
  2) Round the number to three decimal places.

B) Set up a point chart and fill in the numbers you know.

Corner	Distance up or down	Distance right or left	Angle	Distance from point 0
1				
2				
3				
4				
.				
.				
.				

C) Calculate the angle and distance from point 0 to each of the other corners by repeating the following sequence for each corner.
  1) Calculate the angle.
     (Distance up or down) x⇄y **(distance right or left)** INV P→R **(angle in decimal degrees)**
     _____ x⇄y _____ INV P→R _____
  2) Convert from decimal degrees to degrees, minutes, and seconds.
     (Decimal degrees) INV DMS-DD **(degrees, minutes, seconds)**
     _____ INV DMS-DD _____
  3) Find the distance from point 0 to the point.
     x⇄y **(distance in decimal ft.)**
     x⇄y _____
  4) Round the distance to three decimal places.

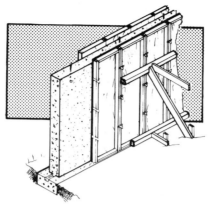

10
Foundation Layout in the Field

Suppose that you are laying out the foundation shown in Figure 9-9, and also suppose that you have already calculated this point chart:

Corner	Distance up or down	Distance right or left	Angle	Distance from point 0
1	0	8'	90°	8'
2	−4.25'	8'	117°58'46"	9.059'
3	−4.25'	106'	92°17'46"	106.085'
4	14.0625'	106'	82°26'35"	106.929'
5	14.0625'	−98'	278° 9'57"	99.004'
6	−4.25'	−98'	267°31'	98.092'
7	−4.25'	0	180°	4.25'

Point 0 has been chosen because it is centrally located and has an unobstructed view of all of the other points. How would you proceed to lay out the points?

Laying Out the Points

Begin by situating the structure on the lot. Put a stake in the ground

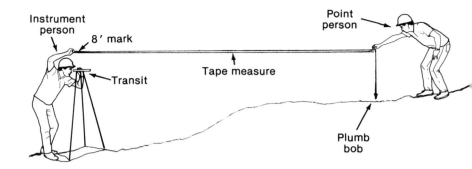

Shooting in point one
Figure 10-1

for point 0 and determine a direction toward corner 1, at point 1. This is normally something the owner will tell you. Then set the transit up over point 0 and level it. Turn the transit toward corner 1, and calibrate the instrument by setting the degree reading exactly at 90° (found in the angle column for point 1).

Now you can start putting the corner stakes in the ground. This requires two people: one behind the instrument, called the *instrument person,* and one finding the points in the ground, called the *point person.* Let's assume you are behind the instrument and a co-worker is finding the points.

To find corner 1, the point person stands so that the instrument person—you—can see your co-worker through the eyepiece. The point person stands 8' (found in the column "distance from point 0" for point 1) from the center of the eyepiece. This distance is found by holding the end of the tape measure while the instrument person lines up the 8' mark with the mark at the center on the top or side of the eyepiece. The tape measure should be roughly level. (Figure 10-1).

Now, you look through the eyepiece and sight the point person. Suppose you sight the point person's belt buckle. Then the point person again holds the end of the tape while you line up the 8' mark with the mark on the instrument. The point person holds the tape at belt-buckle height and drops a plumb bob from the end of the tape into the ground. This is done twice—once about 3" to the right of the belt buckle and again about 3" to the left. The point person lays a straightedge across the centers of both holes (Figure 10-2). The point person then drops the tape and holds a plumb bob by its string with its point about at the center of the straightedge.

Foundation Layout in the Field **169**

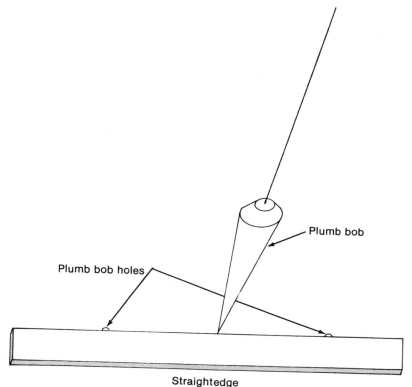

Finding the corner
Figure 10-2

You then look through the instrument and sight the string of the plumb bob. You tell the point person to move right or left until the plumb line lines up with the cross hairs of the instrument. When that happens, the plumb bob is pointing at corner 1.

This is the procedure you will use to shoot in every corner: Turn the transit to the angle on the chart and pull the tape measure to the proper distance. This process works perfectly until you are on a slope and can't see the point person when the transit is level (Figure 10-3).

Suppose you are shooting corner 3. You turn the instrument to 92°17′46″ and find you must tilt the instrument up in order to see the point person. You sight your helper's right knee and look at the instrument. It measures an angle of 16° up from level (Figure 10-4). If you measure the distance at this angle instead of on the level, the distance will be longer than 106.085′. To convert the level distance to

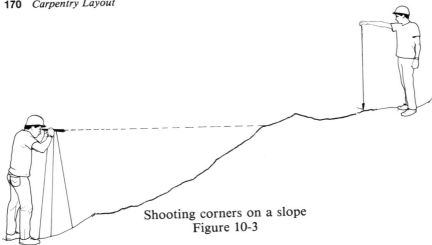

Shooting corners on a slope
Figure 10-3

the distance at 16°, divide 106.085 by the cosine of 16° on your calculator:

106.085 ÷ 16 [cos] = 110.36017

You can then lay out corner 3 in the usual way using 110.360 ′ as the distance.

The rule to follow is this:

(Level distance) ÷ **(angle up or down from level)** [cos] = **(slope distance)**

This rule works whether the angle is up or down from level. Consider some examples involving the other corners.

You are shooting corner 4. You turn the instrument to 82°26′35″ and must tilt the instrument down 27° in order to see the point person. What is the slope distance? (See Figure 10-5).

(Level distance) ÷ **(angle)** [cos] = **(slope distance)**
106.929 ÷ 27 [cos] = 120.00922

So, use a slope distance of 120.009 ′.

For corner 5, you turn the instrument to 278°9′57″ and must tilt the instrument down 47°. What is the slope distance?

(Level distance) ÷ **(angle)** [cos] = **(slope distance)**
99.004 ÷ 47 [cos] = 145.1675

Foundation Layout in the Field 171

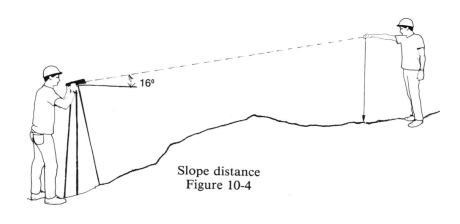

Slope distance
Figure 10-4

Use a slope distance of 145.168'.
You turn the instrument to 267°31' for corner 6. You must raise the instrument 59° to see the point person. What is the slope distance?

(Level distance) ÷ **(angle)** [cos] = **(slope distance)**
98.092 ÷ 59 [cos] = 190.45582

Use a slope distance of 190.456'.
Turn the transit to 180° for corner 7 and tilt it down 42° to see the point person's foot. What is the slope distance?

(Level distance) ÷ **(angle)** [cos] = **(slope distance)**
4.25 ÷ 42 [cos] = 5.7189391

Use a slope distance of 5.719'.
After you have shot all the corners in the ground and put stakes in, you should create references for three corners: point 0, corner 1, and any other corner. With references, you can find these points again after the footings have been dug and poured with concrete.

To create references for these corners, measure back from the corners to two trees or rocks for each corner that won't be disturbed (Figure 10-6). Then, when you are ready to shoot the foundation corners in again, you can measure back to find these three points. You can set the transit up on point 0, calibrate to point 1, and shoot in the rest of the points on the footings so that the forms for the foundation walls will be accurately placed.

172 *Carpentry Layout*

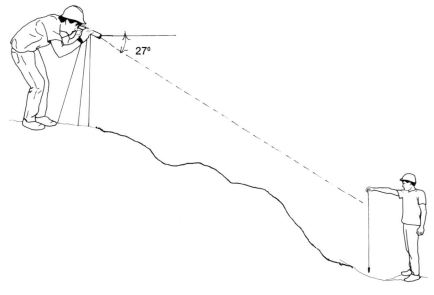

Figuring slope distances
Figure 10-5

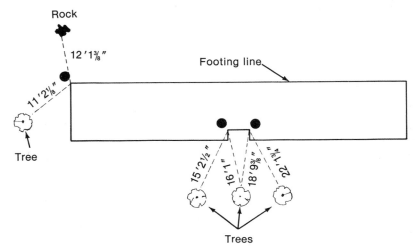

Setting reference points
Figure 10-6

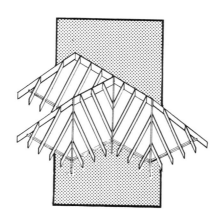

11
Calculating Hip, Valley, and Jack Rafters for Equal-Pitch Roofs

The ability to lay out and cut hip, valley, and jack rafters distinguishes a good carpenter from an average one. Often, carpenters are confused by the geometry involved in these layouts and have difficulty envisioning exactly how the rafter will look. This chapter offers a simple, reliable method for laying out hip, valley, and jack rafters for equal-pitch roofs.

Equal-pitch hip or valley roofs slope in three or four directions instead of two; each roof is of the same pitch. This situation arises with a hip roof and with intersecting roofs that cover a T-or L-shaped plan (Figure 11-1). Each roof section is at a 90° angle to the adjacent section. As a result, the hip or valley rafters are at 45° angles to the roofs on both sides.

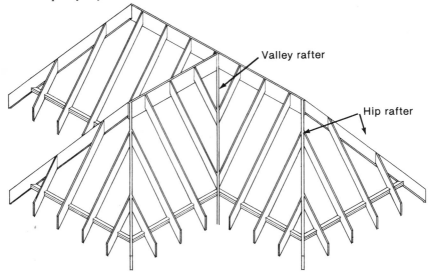

Hip and valley roof
Figure 11-1

Hip Rafters

You begin this chapter by examining hip rafters. You will learn to calculate the following layout dimensions for hip rafters.
- The *total length of the hip rafter* (Figure 11-2).
- The *length from the ridge cut to the birdsmouth plumb cut* (Figure 11-2).
- The *unit rise of the side cuts* of the hip rafter. The side cuts of hip rafters are cuts made so that the hip rafters will fit tightly on the ridge pole, around the corner of the building, or on the facia (Figure 11-2). This unit rise will be used with a unit run of 17, not 12.
- The *depth of the birdsmouth* of the hip rafter (Figure 11-2).
- The pitch of the hip rafter. The pitch of the hip rafter has the same unit rise as the common rafter, but the unit run changes from 12 to approximately 17. Thus on a $5/12$ pitch hip roof, the hip-rafter plumb cut will be $5/17$. This is $1/32$ " off in 17 ", but is accurate enough to use in cutting the plumb and level cuts. You use 17 instead of 12 because you are working at 45° to the roof, and the diagonal of a 12 " square is 16.970563 ", or approximately 17 " (Figure 11-3).

To calculate the layout dimensions for a hip rafter, you need to know the following:
- The *unit rise* of the roof. (Recall that this is the top number given

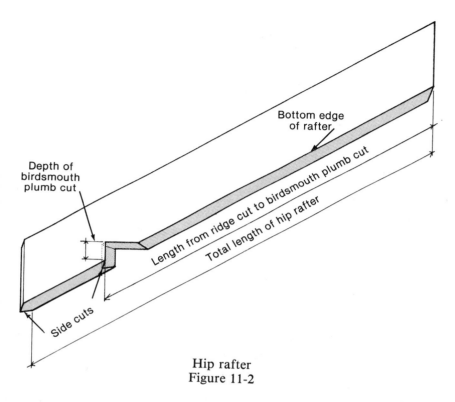

Hip rafter
Figure 11-2

in the roof-pitch ratio. In a 5/12 pitch roof, the unit rise is 5.)
- The *total run* of the common rafter (Figure 5-4). This can be found by measuring directly on the subfloor (as in Chapter 5).
- The *run from the ridge cut to the birdsmouth plumb cut of the common rafter*. This measurement should also be available from earlier calculations.
- *The plumb distance above the birdsmouth of the common rafter* (Figure 11-4). This can be measured directly from a common rafter: Extend the line for the birdsmouth plumb cut to the top of the rafter and measure the distance along that line from the top of the rafter to the seat cut of the birdsmouth.
- The *plumb depth of the hip rafter*. This can be found by placing a framing square on the hip rafter stock and setting it at the *unit rise*/17. Draw a line along the leg of the framing square on which the unit rise is given, extending the line from one edge of the rafter to the other. Measure this line for the plumb depth of the hip rafter (Figure 11-5).

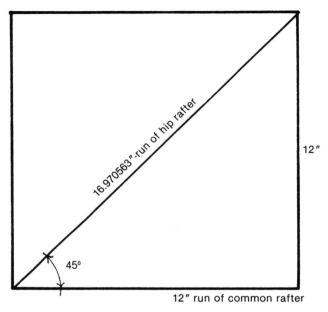

Diagonal of 12″ square
Figure 11-3

Beginning Work

Once you have these facts, you are ready to calculate. First, you find a number, the **diagonal multiplier,** which you multiply by the total run of the common rafter to get the total length of the hip rafter. Then, you multiply the diagonal multiplier by the run of the common rafter from the ridge to the birdsmouth to get the length of the hip rafter from the ridge to the birdsmouth. To find the side-cut unit rise, calculate the rake multiplier as if for a common rafter, except use 17 as the unit run. To find the depth of the birdsmouth plumb cut, subtract the plumb distance above the birdsmouth of the common rafter from the plumb depth of the hip rafter.

On a calculator, proceed like this:

A) First, find the diagonal multiplier.
   1) (Unit rise) x^2 + 12 x^2 + 12 x^2 = $\sqrt{x}$ ÷ 12 = **(diagonal multiplier)**

B) Then, find the total length of the hip rafter in inches.
   1) Convert the run of the common rafter to decimal inches.

Calculating Hip, Valley, and Jack Rafters for Equal-Pitch Roofs

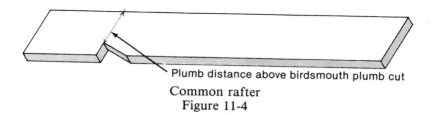

Plumb distance above birdsmouth plumb cut
Common rafter
Figure 11-4

2) **(Total run of common rafter)** ⊠ **(diagonal multiplier)** ⊟ **(length of hip rafter)**
3) Convert the length to fractional inches.

C) Next, find the length of the hip rafter from the ridge cut to the birdsmouth plumb cut.
  1) Convert the run of common rafter from the ridge to the birdsmouth plumb cut to decimal inches.
  2) **(Run of common rafter from ridge cut to birdsmouth plumb cut)** ⊠ **(diagonal multiplier)** ⊟ **(length of hip rafter from the ridge cut to the birdsmouth)**
  3) Convert the length to fractional inches.

D) Find the side-cut unit rise.
  1) **(Unit rise)** x^2 ⊞ 17 x^2 ⊟ $\sqrt{x}$ **(side-cut unit rise)**
  2) Convert the rise to fractional inches.

E) Find the depth of the birdsmouth plumb cut.
  1) Convert the plumb depth to decimal inches.
  2) **(Depth of hip rafter)** ⊟ **(plumb distance above common-rafter birdsmouth)** ⊟ **(depth of birdsmouth plumb cut)**
  3) Convert the depth to fractional inches.

Applications

Example 1
You are framing a $7/12$ equal-pitch hip roof (Figure 11-6). The total run of the common rafters is 12'6", and the run from the plumb cut of the birdsmouth to the ridge is 11'2¼". By measuring a common rafter, you find the plumb distance above the birdsmouth is 8⅜" (Figure 11-7). You draw a line at $7/17$ across the hip rafter, and it measures 12⁷⁄₁₆". Find the dimensions necessary to lay out the hip rafter.

178 *Carpentry Layout*

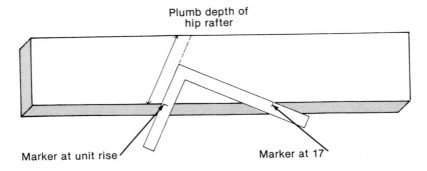

Figuring the plumb depth of a hip rafter
Figure 11-5

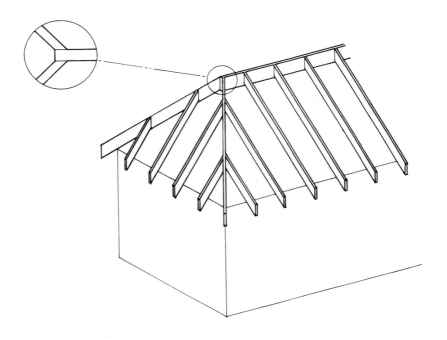

Equal-pitch hip roof
Figure 11-6

Calculating Hip, Valley, and Jack Rafters for Equal-Pitch Roofs **179**

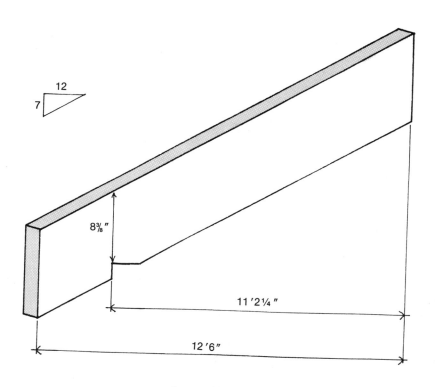

Common rafter
Figure 11-7

Here is what you know:
- The roof pitch is $7/12$; therefore, the unit rise is 7.
- The total run of the common rafter is 12'6".
- The run from the ridge to the birdsmouth plumb cut is 11'2¼".
- The plumb distance above the common-rafter birdsmouth is 8⅜".
- The plumb depth of the hip rafter is 12⁷⁄₁₆".

Now, you are ready to calculate.

A) First, find the diagonal multiplier.
   1) (Unit rise) $\boxed{x^2}$ $\boxed{+}$ 12 $\boxed{x^2}$ $\boxed{+}$ 12 $\boxed{x^2}$ $\boxed{=}$ $\boxed{\sqrt{x}}$ $\boxed{\div}$ 12 $\boxed{=}$ **(diagonal multiplier)**
       7        $\boxed{x^2}$ $\boxed{+}$ 12 $\boxed{x^2}$ $\boxed{+}$ 12 $\boxed{x^2}$ $\boxed{=}$ $\boxed{\sqrt{x}}$ $\boxed{\div}$ 12 $\boxed{=}$ 1.5297966

B) Now find the total length of the hip rafter.

1) Convert the run of the common rafter to decimal inches.
   12 ⊠ 12 ⊞ 6 ⊟ 150
2) **(Total run of common rafter)** ⊠ **(diagonal multiplier)** ⊟ **(length of hip rafter)**
   150 ⊠ 1.5297966 ⊟ 229.46948
3) Convert the length to fractional inches: 229.46948" = 229½".

C) Find the length of the hip rafter from the ridge cut to the birdsmouth plumb cut.
   1) Convert the run of the common rafter from ridge cut to birdsmouth plumb cut to decimal inches.
      11 ⊠ 12 ⊞ 2 ⊞ 1 ⊡ 4 ⊟ 134.25
   2) **(Run of common rafter from ridge cut to birdsmouth plumb cut)** ⊠ **(diagonal multiplier)** ⊟ **(length of hip rafter from the ridge cut to the birdsmouth plumb cut)**
      134.25 ⊠ 1.5297966 ⊟ 205.37519
   3) Convert the length to fractional inches: 205.37519" = 205⅜".

D) Find the side-cut unit rise.
   1) **(Unit rise)** x^2 ⊞ 17 x^2 ⊟ $\sqrt{x}$ **(side-cut unit rise)**
      7   x^2 ⊞ 17 x^2 ⊟ $\sqrt{x}$ 18.384776
   2) Convert the rise to fractional inches: 18.384776" = 18⅜".

E) Find the depth of the birdsmouth plumb cut.
   1) Convert the plumb depth to decimal inches: 12⁷⁄₁₆" = 12.4375".
   2) Convert the plumb distance above the common-rafter birdsmouth to decimal inches: 8⅜" = 8.375".
   3) **(Plumb depth of hip rafter)** ⊟ **(plumb distance above common-rafter birdsmouth)** ⊟ **(depth of birdsmouth)**
      12.4375 ⊟ 8.375 ⊟ 4.0625
   4) Convert the depth to fractional inches: 4.0625" = 4¹⁄₁₆".

Before you begin work, summarize your answers (Figure 11-8).
- Length of the hip rafter = 229½"
- Distance to birdsmouth = 205⅜"
- Side cut = 18⅜/17
- Depth of birdsmouth = 4¹⁄₁₆"
- Plumb cut = ⁷⁄₁₇

Example 2
You're framing a hip roof with a ¹⁴⁄₁₂ pitch. The total common rafter run is 16'3½", and the run from the ridge cut to the birdsmouth plumb cut is 14'9½" (Figure 11-9). You find the plumb distance

Calculating Hip, Valley, and Jack Rafters for Equal-Pitch Roofs   181

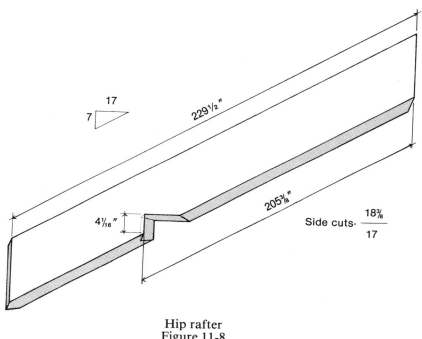

Hip rafter
Figure 11-8

above the birdsmouth to be 4⅛". You draw a line at ¹⁴⁄₁₇ across the hip-rafter stock, and it measures 7⅛". Find all the dimensions necessary to lay out the hip rafter.

Here is what you know:
- The roof pitch is ¹⁴⁄₁₂; therefore, the unit rise is 14.
- The total run of the common rafter is 16'3½".
- The run from the ridge to the birdsmouth plumb cut is 14'9½".
- The plumb distance above the common-rafter birdsmouth is 4⅛".
- The plumb depth of the hip rafter is 7⅛".

Now, you are ready to calculate.

A) Find the diagonal multiplier.
   1) **(Unit rise)** x^2 $+$ 12 x^2 $+$ 12 x^2 $=$ $\sqrt{x}$ $\div$ 12 $=$ **(diagonal multiplier)**
      14   x^2 $+$ 12 x^2 $+$ 12 x^2 $=$ $\sqrt{x}$ $\div$ 12 $=$ 1.8333333

B) Now find the total length of the hip rafter.

182 *Carpentry Layout*

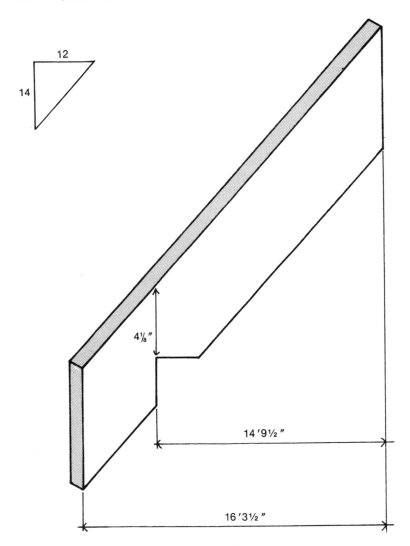

Common rafter
Figure 11-9

1) Convert the run of the common rafter to decimal inches.
16 ⊠ 12 ⊞ 3 ⊞ 1 ⊟ 2 ⊟ 195.5
2) **(Total run of common rafter)** ⊠ **(diagonal multiplier)** ⊟ **(length of hip rafter)**
195.5 ⊠ 1.8333333 ⊟ 358.416667
3) Convert the length to fractional inches: 358.416667″ = 358 $\frac{7}{16}$″.

C) Find the length of the hip rafter from the ridge cut to the birdsmouth plumb cut.
1) Convert the run of the common rafter from the ridge cut to the birdsmouth plumb cut to decimal inches.
14 ⊠ 12 ⊞ 9 ⊞ 1 ⊟ 2 ⊟ 177.5
2) **(Run of common rafter from ridge cut to birdsmouth plumb cut)** ⊠ **(diagonal multiplier)** ⊟ **(length of hip rafter from the ridge cut to the birdsmouth plumb cut)**
177.5 ⊠ 1.8333333 ⊟ 325.41666
3) Convert the length to fractional inches: 325.41666″ = 325 $\frac{7}{16}$″.

D) Find the side-cut unit rise.
1) **(Unit rise)** x^2 ⊞ 17 x^2 ⊟ $\sqrt{x}$ **(side-cut unit rise)**
14 x^2 ⊞ 17 x^2 ⊟ $\sqrt{x}$ 22.022716
2) Convert the rise to fractional inches: 22.022716″ = 22″.

E) Find the depth of the birdsmouth plumb cut.
1) Convert the plumb depth of the hip rafter to decimal inches: 7$\frac{1}{8}$″ = 7.125″.
2) Convert the plumb depth above the common-rafter birdsmouth to decimal inches: 4$\frac{1}{8}$″ = 4.125″.
3) **(Plumb depth of hip rafter)** ⊟ **(plumb distance above common-rafter birdsmouth)** ⊟ **(depth of birdsmouth)**
7.125 ⊟ 4.125 ⊟ 3

Summarize your answers (Figure 11-10).
- Length of the hip rafter = 358 $\frac{7}{16}$″
- Distance to birdsmouth = 325 $\frac{7}{16}$″
- Side cut = $\frac{22}{17}$
- Depth of birdsmouth = 3″
- Plumb cut = $\frac{14}{17}$

Valley Rafters

The process for laying out a valley rafter is the same as that for laying out a hip rafter with one deviation: The depth of the plumb cut is

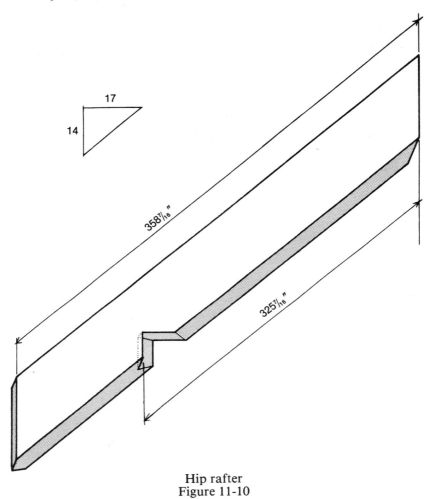

Hip rafter
Figure 11-10

figured differently. If you cut the birdsmouth to the same depth as you did the hip rafter, the roof sheathing won't lie flat over the valley rafter (Figure 11-11). To remedy this, you must "drop" the rafter by deepening the birdsmouth plumb cut. The amount to drop the rafter is equal to the unit rise of the roof divided by 17 and multiplied by half the thickness of the valley rafter. On the calculator, you would do these steps:

(Unit rise) ÷ 17 × .5 × **(thickness of valley rafter)** = **(drop)**

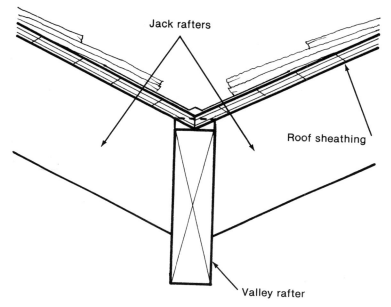

Dropping the valley rafter
Figure 11-11

You add this drop to the depth of the birdsmouth that you figure as with hip rafters to get a total birdsmouth depth. So, in addition to knowing the five dimensions for hip rafters, you must also know the thickness of the valley rafter.

Applications

Consider some examples.

Example 3
You are framing a structure with a 5/12 equal-pitch intersecting roof (Figure 11-12). The valley rafter is a 6 × 12. The run of the common rafter is 18′3″, and the run from the plumb cut of the birdsmouth to the ridge is 17′1″. By measuring a common rafter, you find the plumb distance above the birdsmouth to be 10⅛″. You draw a line at 5/17 across the valley rafter and it measures 12″. Find the dimensions necessary to lay out the valley rafter.

Here is what you know:
- The roof pitch is 5/12; therefore, the unit rise is 5.

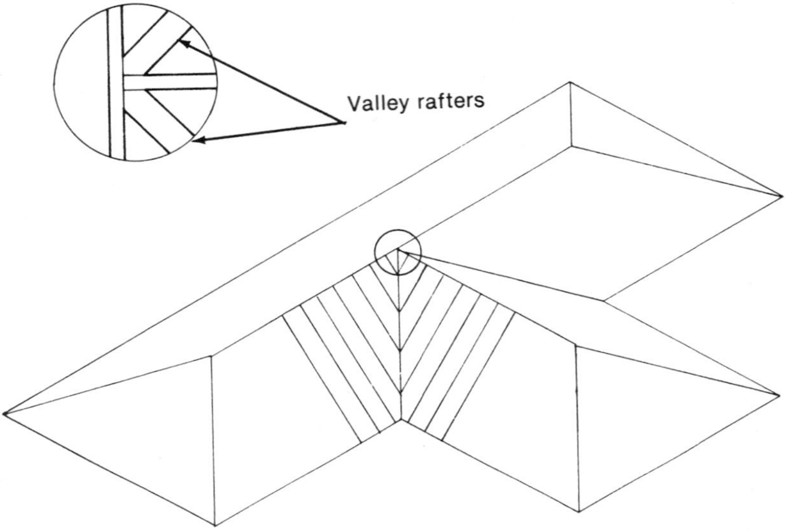

Equal-pitch roof with valley rafters
Figure 11-12

- The total run of the common rafter is 18′3″.
- The run from the ridge to the birdsmouth plumb cut is 17′1″.
- The plumb distance above the common-rafter birdsmouth is 10⅛″.
- The plumb depth of the valley rafter is 12″.
- The width of the 6 × 12 valley rafter is 5½″.

Now, you are ready to calculate.

A) First, find the diagonal multiplier.
   1) **(Unit rise)** x^2 + 12 x^2 + 12 x^2 = $\sqrt{x}$ ÷ 12 = **(diagonal multiplier)**
      5 x^2 + 12 x^2 + 12 x^2 = $\sqrt{x}$ ÷ 12 = 1.4743172

B) Next, find the total length of the valley rafter.
   1) Convert the run of the common rafter to decimal inches.
      18 × 12 + 3 = 219
   2) **(Total run of common rafter)** × **(diagonal multiplier)** = **(length of valley rafter)**
      219 × 1.4743172 = 322.87547
   3) Convert the length to fractional inches: 322.87547″ = 322⅞″.

Calculating Hip, Valley, and Jack Rafters for Equal-Pitch Roofs

C) Then, find the length of the valley rafter from the ridge cut to the birdsmouth plumb cut.
   1) Convert the run of the common rafter from the ridge cut to the birdsmouth plumb cut to decimal inches.
      17 ⊠ 12 ⊞ 1 ⊟ 205
   2) **(Run of common rafter from ridge cut to birdsmouth plumb cut) ⊠ (diagonal multiplier) ⊟ (length of valley rafter from the ridge cut to the birdsmouth plumb cut)**
      205 ⊠ 1.4743172 ⊟ 302.23503
   3) Convert the length to fractional inches: 302.23503″ = 302¼″.

D) Now, find the side-cut unit rise.
   1) **(Unit rise)** x^2 ⊞ 17 x^2 ⊟ $\sqrt{x}$ **(side-cut unit rise)**
      5   x^2 ⊞ 17 x^2 ⊟ $\sqrt{x}$ 17.720045
   2) Convert the rise to fractional inches: 17.720045″ = 17¾″.

E) Finally, find the depth of the birdsmouth plumb cut.
   1) Convert the plumb depth of the valley rafter to decimal inches: 12″ = 12″.
   2) Convert the distance above the common-rafter birdsmouth to decimal inches: 10⅛″ = 10.125″.
   3) **(Plumb depth of valley rafter) ⊟ (plumb distance above common-rafter birdsmouth) ⊟ (depth of birdsmouth)**
      12 ⊟ 10.125 ⊟ 1.875
   4) Convert the thickness of the valley rafter to decimal inches: 5½″ = 5.5″.
   5) Find the drop of the valley rafter.
     (Unit rise) ⊟ 17 ⊠ .5 ⊠ (thickness of valley rafter) ⊟ (drop)
      5   ⊟ 17 ⊠ .5 ⊠   5.5   ⊟ .8088235
   6) Add the drop to the birdsmouth depth.
      .8088235 ⊞ 1.875 ⊟ 2.6838235
   7) Convert the depth to fractional inches: 2.6838235 = 2¹¹⁄₁₆″.

  Summarize your answers (Figure 11-13).
- Length of the valley rafter = 322⅞″
- Distance to birdsmouth = 302¼″
- Side cut = 17¾/17
- Total depth of birdsmouth = 2¹¹⁄₁₆″
- Pitch of the valley rafter = ⁵⁄₁₇

Jack Rafters

Jack rafters are common rafters that attach to a hip or valley rafter

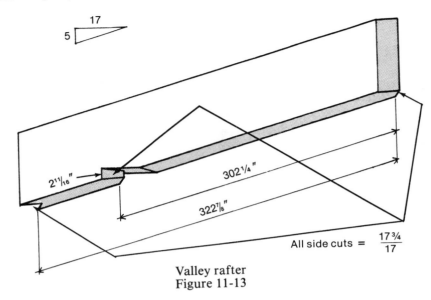

Valley rafter
Figure 11-13

on one end (Figure 11-14). Jack rafters have a side cut as well as a plumb cut where they attach to the hip or valley rafters, and each jack rafter is shorter than the one before it by a common difference (Figure 11-14). You will learn a method of finding the side-cut unit rise (use this with a unit run of 12 to lay out the side cut on the jack) and the length of each jack rafter. The plumb and level cuts are the same as for the common rafters.

Beginning Work

To calculate the side-cut unit rise and the jack lengths, you must know the following:
- The *roof-pitch unit rise.*
- The *spacing of the jacks.* This is the same spacing as the common rafters (Figure 11-14).
- The *length of the common rafters.*
- The *distance from the intersection of the hip or valley rafter and the ridge pole to the nearest common rafter* (Figure 11-14). This is called the **distance to the nearest common rafter** from the hip or valley rafter.

Find the unit rise of the side cut by multiplying the rake multiplier by 12. Then, find the common difference of the jack rafters. By

Calculating Hip, Valley, and Jack Rafters for Equal-Pitch Roofs

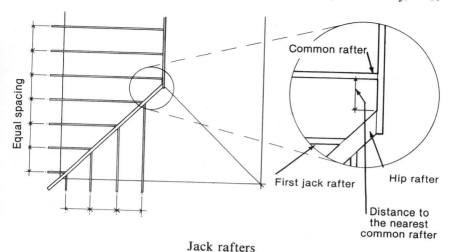

Jack rafters
Figure 11-14

subtracting the common difference from the length of one jack, you get the length of the next one. You find the **common difference** by multiplying the rake multiplier by the spacing. You then calculate the length of the first jack rafter by subtracting the distance from the nearest common rafter to the hip or valley rafter from the spacing of the jacks and multiplying that answer by the rake multiplier. Then, you subtract that distance from the common rafter length to get the length of the first jack rafter. To find the length of the second jack rafter, you subtract the common difference from the length of the first jack rafter. The length of the third jack rafter is found by subtracting the common difference from the length of the second jack rafter, and so on, until all the jack-rafter lengths are found.

On a calculator, proceed as follows:
A) Find the rake multiplier.
  1) (Unit rise) x^2 $+$ 12 x^2 $=$ $\sqrt{x}$ $\div$ 12 $=$ **(rake multiplier)**

B) Find the unit rise of the side cut.
  1) **(Rake multiplier)** $\times$ 12 $=$ **(unit rise of side cut)**
  2) Convert the rise to fractional inches.

C) Find the common difference.
  1) **(Rake multiplier)** $\times$ **(spacing)** $=$ **(common difference)**

D) Find the first-jack length.
  1) Convert the length of the common rafter to decimal inches.
  2) **(Spacing)** ⊟ **(distance to nearest common rafter from hip or valley)** ⊟ ⊠ **(rake multiplier)** ⊟ +/− ⊞ **(common-rafter length)** ⊟ **(length of first jack rafter)**
  3) Convert the length to fractional inches.

E) Find the remaining jack lengths:
  1) **(Preceding-jack length)** ⊟ **(common difference)** ⊟ **(jack length)**
  2) Convert the length to fractional inches.

Applications

Consider some examples.

Example 4
You are working on a $10/12$ pitch hip roof. The common rafters and hip rafters are in place, and you are ready to frame the jack rafters. The rafter spacing is 16″ on center, and the nearest common rafter is 8¼″ from the intersection of the hip rafter with the ridge pole (Figure 11-15). The common rafters are 14′3⅜″ long. Find the lengths of all the jack rafters and the side cut required.
  Here is what you know:
- The roof pitch is $10/12$.
- The spacing is 16″.
- The length of the common rafter is 14′3⅜″.
- The distance to the nearest common rafter from the hip rafter is 8¼″.

Now you are ready to calculate.

A) Find the rake multiplier.
  1) **(Unit rise)** x^2 ⊞ 12 x^2 ⊟ $\sqrt{x}$ ⊟ 12 ⊟ **(rake multiplier)**
     10 x^2 ⊞ 12 x^2 ⊟ $\sqrt{x}$ ⊟ 12 ⊟ 1.3017083
B) Find the unit rise of the side cut.
  1) **(Rake multiplier)** ⊠ 12 ⊟ **(unit rise of side cut)**
     1.3017083 ⊠ 12 ⊟ 15.620499
  2) Convert the rise to fractional inches: 15.620499″ = 15⅝″.

C) Find the common difference.
  1) **(Rake multiplier)** ⊠ **(spacing)** ⊟ **(common difference)**
     1.3017083   ⊠   16   ⊟ 20.827333

D) Find the first-jack length.

Calculating Hip, Valley, and Jack Rafters for Equal-Pitch Roofs **191**

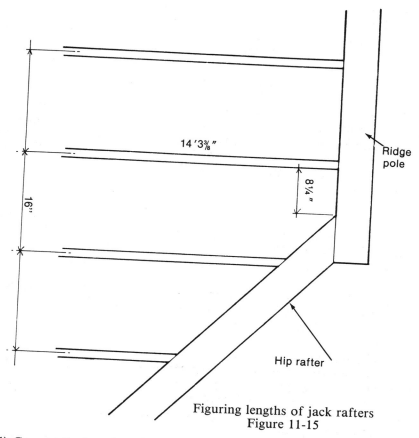

Figuring lengths of jack rafters
Figure 11-15

1) Convert the length of the common rafter to decimal inches.
14 ⊠ 12 ⊞ 3 ⊞ 3 ⊡ 8 ⊟ 171.375

2) (Spacing) ⊟ **(distance to nearest common rafter from hip rafter)** ⊟ ⊠ **(rake multiplier)** ⊟ ⊞ ⊞ **(common-rafter length)** ⊟ **(length of first jack rafter)**
16 ⊟ 8.25 ⊟ ⊠ 1.3017083 ⊟ ⊞ ⊞ 171.375 ⊟ 161.28676
Convert the length to fractional inches: 161.28676 = 161 5/16″.
The ⊞ key changes the sign of the number on the display so that it can be subtracted from the common-rafter length without being re-entered on the calculator.

E) Find the remaining jack lengths.
First, find the second-jack length.
1) (First-jack length) ⊟ **(common difference)** ⊟ **(second-jack length)**
    161.28676   ⊟   20.827333   ⊟ 140.45943

2) Convert the length to fractional inches: 140.45943″ = 140 7/16″.

Find the third-jack length.
1) **(Second-jack length)** ⊟ **(common difference)** ⊟ **(third-jack length)**
   140.45943   ⊟   20.827333   ⊟   119.6321
2) Convert the length to fractional inches: 119.6321″ = 119 5/8″.

Find the fourth-jack length.
1) **(Third-jack length)** ⊟ **(common difference)** ⊟ **(fourth-jack length)**
   119.6321   ⊟   20.827333   ⊟   98.804766
2) Convert the length to fractional inches: 98.804766″ = 98 13/16″.

Find the fifth-jack length.
1) **(Fourth-jack length)** ⊟ **(common difference)** ⊟ **(fifth-jack length)**
   98.804766   ⊟   20.827333   ⊟   77.977434
2) Convert the length to fractional inches: 77.977434″ = 78″.

Find the sixth-jack length.
1) **(Fifth-jack length)** ⊟ **(common difference)** ⊟ **(sixth-jack length)**
   77.977434   ⊟   20.827333   ⊟   57.150102
2) Convert the length to fractional inches: 57.150102″ = 57 1/8″.

Find the seventh-jack length.
1) **(Sixth-jack length)** ⊟ **(common difference)** ⊟ **(seventh-jack length)**
   57.150102   ⊟   20.827333   ⊟   36.32277
2) Convert the length to fractional inches: 36.32277″ = 36 5/16″.

Find the eighth-jack length.
1) **(Seventh-jack length)** ⊟ **(common difference)** ⊟ **(eighth-jack length)**
   36.32277   ⊟   20.827333   ⊟   15.495438
2) Convert the length to fractional inches: 15.495438″ = 15 1/2″.

Before you begin work, summarize your answers:
- Side cut = 15 5/8 /12 (Figure 11-16)
- Jack-rafter lengths to the short side = 161 5/16″, 140 7/16″, 119 5/8″, 98 13/16″, 78″, 57 1/8″, 36 5/16″, and 15 1/2″

Example 5
You are working on a 9/12 intersecting roof with the common rafters and valley rafters in place. The rafter spacing is 24″ on center, and the nearest common rafter is 16 1/2″ from the tail of the valley rafter (Figure 11-17). The common rafters are 9′6″ long. Find the length of

Calculating Hip, Valley, and Jack Rafters for Equal-Pitch Roofs **193**

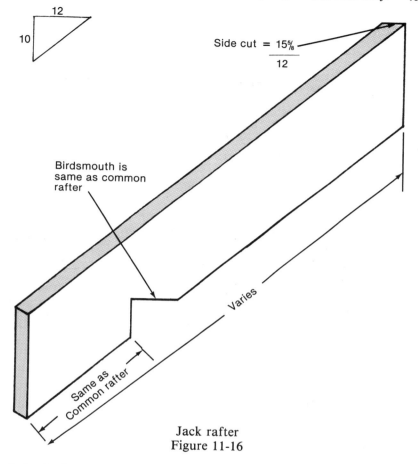

Jack rafter
Figure 11-16

all the jack rafters and the side cut required.
   Here is what you know:
- The roof pitch is 9/12.
- The spacing is 24".
- The length of the common rafter is 9'6".
- The distance to the nearest common rafter from the hip rafter is 16½".

Now you are ready to calculate.
A) Find the rake multiplier.

   (Unit rise) $\boxed{x^2}$ $\boxed{+}$ 12 $\boxed{x^2}$ $\boxed{=}$ $\boxed{\sqrt{x}}$ $\boxed{\div}$ 12 $\boxed{=}$ **(rake multiplier)**
   9       $\boxed{x^2}$ $\boxed{+}$ 12 $\boxed{x^2}$ $\boxed{=}$ $\boxed{\sqrt{x}}$ $\boxed{\div}$ 12 $\boxed{=}$ 1.25

194 Carpentry Layout

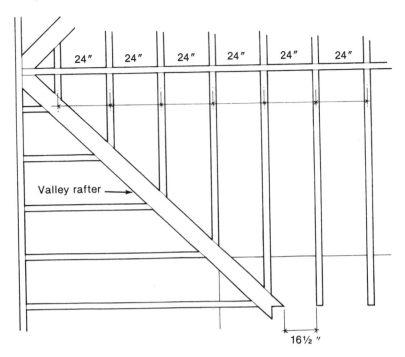

Plan view of jack rafters with valley rafters
Figure 11-17

B) Find the unit rise of the side cut.
  1) **(Rake multiplier)** ☒ 12 ▭ **(unit rise of side cut)**
     1.25   ☒ 12 ▭ 15
  2) The unit rise is 15″.

C) Find the common difference.
  (Rake multiplier) ☒ **(spacing)** ▭ **(common difference)**
     1.25   ☒   24   ▭ 30

D) Find the first-jack length.
  1) Convert the length of the common rafter to decimal inches.
     9 ☒ 12 ⊞ 6 ▭ 114
  2) **(Spacing)** ⊟ **(distance to nearest common rafter from valley rafter)** ▭ ☒ **(rake multiplier)** ▭ ± ⊞ **(common-rafter length)** ▭ **(length of first jack rafter)**
     24 ⊟ 16.5 ▭ ☒ 1.25 ▭ ± ⊞ 114 ▭ 104.625
  3) Convert the length to fractional inches: 104.625″ = 104⅝″

E) Find the other jack lengths.
Find the second-jack length.
1) **(First-jack length)** ⊟ **(common difference)** ⊟ **(second-jack length)**
   104.625   ⊟   30   ⊟   74.625
2) Convert the length to fractional inches: 74.625" = 74⅝".

Find the third-jack length.
1) **(Second-jack length)** ⊟ **(common difference)** ⊟ **(third-jack length)**
   74.625   ⊟   30   ⊟   44.625
2) Convert the length to fractional inches: 44.625" = 44⅝".

Find the fourth-jack length.
1) **(Third-jack length)** ⊟ **(common difference)** ⊟ **(fourth-jack length)**
   44.625   ⊟   30   ⊟   14.625
2) Convert the length to fractional inches: 14.625" = 14⅝".

Summarize your answers before you begin work:
- Side cut = $15/12$
- Jack-rafter lengths to the short side = 104⅝", 74⅝", 44⅝", 14⅝"

Problems

1) You are framing a $5/12$ equal-pitch hip roof. The total run of the common rafter is 15′, and the run from the plumb cut of the birdsmouth to the ridge is 13′. By measuring a common rafter, you find the plumb distance above the birdsmouth to be 9¼″. You draw a line at $5/17$ across the hip rafter, and it measures 13⅜″. Find all the dimensions necessary to lay out the hip rafter.

2) You are framing a structure with a 10½/12 equal-pitch intersecting roof. The valley rafter is a 4 × 12. The run of the common rafter is 12′2″, and the run from the plumb cut of the birdsmouth to the ridge is 10′8″. By measuring a common rafter, you find the plumb distance above the birdsmouth to be 11¼″. You draw a line at 10½/17 across the valley rafter, and it measures 13½″. Find the dimensions necessary to lay out the valley rafter.

3) You are working on a $7/12$ pitch hip roof. The common rafters and hip rafters are in place, and you are ready to frame the jack rafters. The rafter spacing is 32″ on center, and the nearest common rafter is 21½″ from the intersection of the hip rafter with the ridge pole. The common rafters are 12′2½″ long. Find the lengths of all the jack rafters and the side cut required.

Worksheet

HIP RAFTER

A) First, find the diagonal multiplier.
 1) **(Unit rise)** x^2 $+$ 12 x^2 $+$ 12 x^2 $=$ $\sqrt{x}$ $\div$ 12 $=$ (diagonal multiplier)
 _____ x^2 $+$ 12 x^2 $+$ 12 x^2 $=$ $\sqrt{x}$ $\div$ 12 $=$ _____

B) Then, find the total length of the hip rafter in inches.
 1) Convert the run of the common rafter to decimal inches.
 2) **(Total run of common rafter)** $\times$ **(diagonal multiplier)** $=$ **(length of hip rafter)**
 _____ $\times$ _____ $=$ _____
 3) Convert the length to fractional inches.

C) Next, find the length of the hip rafter from the ridge cut to the birdsmouth plumb cut.
 1) Convert the run of the common rafter from the ridge to the birdsmouth plumb cut to decimal inches.
 2) **(Run of common rafter from ridge cut to birdsmouth plumb cut)** $\times$ **(diagonal multiplier)** $=$ **(length of hip rafter from the ridge cut to the birdsmouth)**
 _____ $\times$ _____ $=$ _____
 3) Convert the length to fractional inches.

D) Now, find the side-cut unit rise.
 1) **(Unit rise)** x^2 $+$ 17 x^2 $=$ $\sqrt{x}$ **(side-cut unit rise)**
 _____ x^2 $+$ 17 x^2 $=$ $\sqrt{x}$ _____
 2) Convert the rise to fractional inches.

E) Finally, find the depth of the birdsmouth plumb cut.
 1) Convert the plumb depth to decimal inches.
 2) Convert the plumb distance above the common-rafter birdsmouth to decimal inches.
 3) **(Plumb depth of hip rafter)** $-$ **(plumb distance above common-rafter birdsmouth)** $=$ **(depth of birdsmouth plumb cut)**
 _____ $-$ _____ $=$ _____
 4) Convert the depth to fractional inches.

Worksheet

VALLEY RAFTER

A) First, find the diagonal multiplier.
   (Unit rise) x^2 $+$ 12 x^2 $+$ 12 x^2 $=$ $\sqrt{x}$ $\div$ 12 $=$ **(diagonal multiplier)**
   _____ x^2 $+$ 12 x^2 $+$ 12 x^2 $=$ $\sqrt{x}$ $\div$ 12 $=$ _____

B) Then, find the total length of the valley rafter in inches.
   1) Convert the run of the common rafter to decimal inches.
   2) **(Total run of common rafter)** $\times$ **(diagonal multiplier)** $=$ **(length of valley rafter)**
      _____ $\times$ _____ $=$ _____
   3) Convert the length to fractional inches.

C) Next, find the length of the valley rafter from the ridge cut to the birdsmouth plumb cut.
   1) Convert the run of the common rafter from the ridge to the birdsmouth plumb cut to decimal inches.
   2) **(Run of common rafter from ridge cut to birdsmouth plumb cut)** $\times$ **(diagonal multiplier)** $=$ **(length of valley rafter from the ridge cut to the birdsmouth)**
      _____ $\times$ _____ $=$ _____
   3) Convert the length to fractional inches.

D) Now, find the side-cut unit rise.
   1) **(Unit rise)** x^2 $+$ 17 x^2 $=$ $\sqrt{x}$ **(side-cut unit rise)**
      _____ x^2 $+$ 17 x^2 $=$ $\sqrt{x}$ _____
   2) Convert the rise to fractional inches.

E) Finally, find the depth of the birdsmouth plumb cut.
   1) Convert the plumb depth of the valley rafter to decimal inches.
   2) Convert the distance above the common-rafter birdsmouth to decimal inches.
   3) **(Plumb depth of hip rafter)** $-$ **(plumb distance above common-rafter birdsmouth)** $=$ **(depth of birdsmouth)**
      _____ $-$ _____ $=$ _____
   4) Convert the thickness of the valley rafter to inches.
   5) Find the drop of the valley rafter.
      (Unit rise) $\div$ 17 $\times$ .5 $\times$ **(thickness of valley rafter)** $=$ **(drop)**
      _____ $\div$ 17 $\times$ .5 $\times$ _____ $=$ _____

Calculating Hip, Valley, and Jack Rafters for Equal-Pitch Roofs **199**

6) Add the drop to the birdsmouth depth.
_____ $+$ _____ $=$ _____

7) Convert the depth to fractional inches.

Worksheet

JACK RAFTER

A) Find the rake multiplier.
  1) **(Unit rise)** x^2 $+$ 12 x^2 $=$ $\sqrt{x}$ $\div$ 12 $=$ **(rake multiplier)**
  _____ x^2 $+$ 12 x^2 $=$ $\sqrt{x}$ $\div$ 12 $=$ _____

B) Find the unit rise of the side cut.
  1) **(Rake multiplier)** $\times$ 12 $=$ **(unit rise of side cut)**
  _____ $\times$ 12 $=$ _____
  2) Convert the rise to fractional inches.

C) Find the common difference.
  (Rake multiplier) $\times$ **(spacing)** $=$ **(common difference)**
  _____ $\times$ _____ $=$ _____

D) Find the first-jack length.
  1) Convert the length of the common rafter to decimal inches.
  2) **(Spacing)** $-$ **(distance to nearest common rafter from hip or valley)** $=$ $\times$ **(rake multiplier)** $=$ $+/-$ $+$ **(common-rafter length)** $=$ **(length of first jack rafter)**
  _____ $-$ _____ $=$ $\times$ _____ $=$ $+/-$ $+$ _____ $=$ _____
  3) Convert the length to fractional inches.

E) Find the other jack lengths. (Repeat this step until all jack lengths are found.)
  1) **(Preceding-jack length)** $-$ **(common difference)** $=$ **(jack length)**
  _____ $-$ _____ $=$ _____
  2) Convert the length to fractional inches.

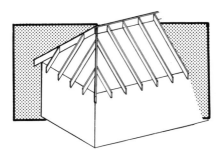

12
Practical Application: Laying Out Hip, Valley, and Jack Rafters for Equal-Pitch Roofs

Once you have calculated the layout dimensions needed for a hip or intersecting roof, you have a good start—but there is still work ahead! Now, you must transfer the layout to the hip, valley, or jack rafter that must be cut. This chapter shows you a practical way to cut hip, valley, and jack rafters.

Laying Out Rafters

To lay out the rafters, you will need the following:
- Rafter square with markers.
- A tape measure. A 25-foot measure is ideal.

- A motorized hand circular saw with an adjustable table (skill saw).
- The calculations from Chapter 11. These include:
  The roof pitch.
  The total length of the rafter.
  The length of the rafter from the ridge cut to the birdsmouth plumb cut.
  The side cut.
  The depth of the birdsmouth.
  The pitch of the hip rafter.

The ability to visualize the rafter in place is very important: If you can visualize how every cut fits into place, you will prevent many costly mistakes.

As an example, suppose you want to lay out the hip rafter that was calculated in Example 1 of Chapter 11.
- Roof pitch = $7/12$.
- Total rafter length = $229\frac{1}{2}$″.
- Length from the ridge cut to the birdsmouth plumb cut = $205\frac{3}{8}$″.
- Side cut = $18\frac{3}{8}/17$.
- Depth of the birdsmouth = $4\frac{1}{16}$″.
- Pitch of the hip rafter = $7/17$.

The first step is to choose a straight piece of lumber to use as the hip rafter. Be sure it is at least a foot longer than the length of the hip rafter. Then draw an arrow pointing to the crown (the humped side); the arrow points to the top of the rafter. This will help you visualize the rafter in place (Figure 12-1).

Next, set the markers on the square. Since this is a hip rafter, set the markers at 7″ on the tongue and 17″ (*not* 12″) on the body. Plumb cuts will be drawn on the tongue (where 7″ is set), and level cuts will be drawn on the body (where 17″ is set). Place the rafter square as shown in Figure 12-2. Draw a plumb line along the tongue. This gives you the ridge cut. It is important to visualize how the hip rafter joins the ridge pole because the length from the ridge cut to the birdsmouth plumb cut will be different on each side (Figure 12-3). The length we have calculated is for the long side.

Now measure along the bottom edge of the rafter (the edge to which the arrow is not pointing) from the ridge cut line, and put marks at $205\frac{3}{8}$″ and $229\frac{1}{2}$″ (Figure 12-4). These marks locate the birdsmouth plumb cut and the tail plumb cut. Slide the rafter square

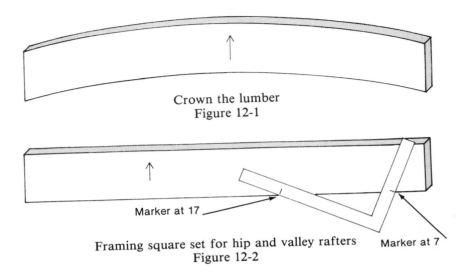

Crown the lumber
Figure 12-1

Framing square set for hip and valley rafters
Figure 12-2

Marker at 17

Marker at 7

to the birdsmouth plumb cut. Mark and draw another plumb line along the tongue of the rafter square (Figure 12-5). Making sure that the markers are tight against the rafter, measure up $4\frac{1}{16}''$ from the bottom of the rafter on this line—the depth of the birdsmouth (Figure 12-6). Slide the rafter square until the body lines up with the mark at $4\frac{1}{16}''$ and draw a line along the body of the square (Figure 12-7). This line is the birdsmouth seat cut.

Now slide the rafter square to the tail cut mark and draw a plumb line along the tongue (Figure 12-8). This is the tail cut line. Again, make sure the markers are tight against the rafter. If the board is too short to accomodate the rafter square, try using an extension as shown.

At this point, you have laid out one side of the rafter. To continue the layout, turn the rafter crown down (Figure 12-9). The arrow on the rafter now points at the ground. On the bottom edge of the rafter, you are working with the side cut, $18\frac{3}{8}/17$. You can't use this proportion as it is because the tongue of the rafter square doesn't go to $17''$; it is only $16''$ long. To find an equal side cut, divide each number in half to get $9\frac{3}{16}/8\frac{1}{2}$.

$$18\frac{3}{8} \div 2 = 9\frac{3}{16}$$
$$17 \div 2 = 8\frac{1}{2}$$

Use this pitch for all side cuts: at the ridge, at the birdsmouth, and at the tail cut.

Practical Application: Laying Out Hip, Valley, and Jack Rafters for Equal-Pitch Roofs 203

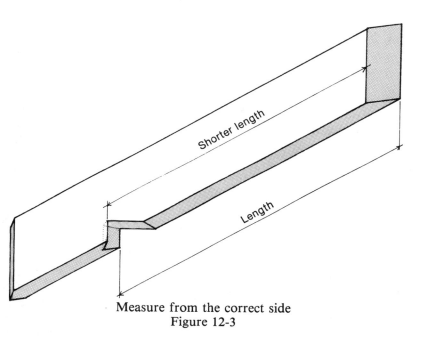

Measure from the correct side
Figure 12-3

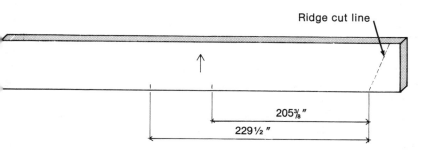

Measure for birdsmouth and tail cut
Figure 12-4

204 Carpentry Layout

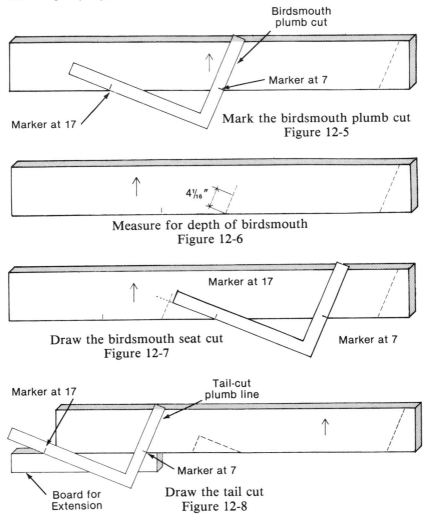

Mark the birdsmouth plumb cut
Figure 12-5

Measure for depth of birdsmouth
Figure 12-6

Draw the birdsmouth seat cut
Figure 12-7

Draw the tail cut
Figure 12-8

The line you draw for the side cut is always along the leg of the square with the first number of the pitch (9 3/16" in this case) on it. Lay out the ridge side cut first by lining up the rafter square with the plumb cut (Figure 12-10) and drawing the side cut.

Now move to the birdsmouth. First, draw a line square across the edge of the rafter from the birdsmouth plumb cut line. Then draw the birdsmouth side cuts (Figure 12-11 & 12-12). These allow the hip rafter to fit tightly around the corner of the building (Figure 12-13).

Practical Application: Laying Out Hip, Valley, and Jack Rafters for Equal-Pitch Roofs

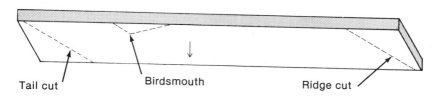

Turn the rafter crown down
Figure 12-9

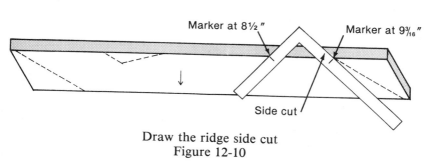

Draw the ridge side cut
Figure 12-10

Next move to the tail side cut. If it is to be exposed, draw a line square across the edge of the hip rafter (Figure 12-14). If it will be covered by facia, it must be cut to receive the facia: Lay out these lines as you did with the birdsmouth side cuts (Figures 12-15, 12-16, & 12-17).

Now roll the hip rafter to the other side (Figure 12-18). The side of the rafter you started with is now facing the ground. Draw another ridge-cut plumb line from the side-cut line to the top of the rafter. Be sure everything is right by checking the other side of the rafter: The lines should slope in the same direction (Figure 12-19).

Next slide the rafter square to the birdsmouth side-cut line and draw the plumb cut (Figure 12-20). Measure up 4 1/16" and draw the seat cut line as on the other side (Figure 12-21).

Slide the rafter square to the tail-cut line and draw the plumb cut (Figure 12-22). This completes the layout; now you are ready to cut.

Begin cutting at the tail cut. If the tail cut does not need to receive the facia, simply cut along the plumb line from either side. If the tail cut is to receive the facia, set the skill saw table at 45° and cut along

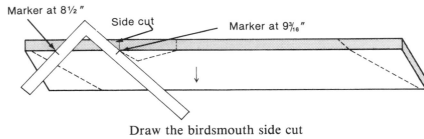

Draw the birdsmouth side cut
Figure 12-11

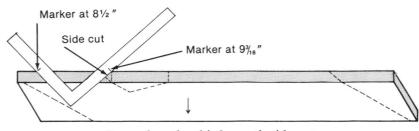

Draw the other birdsmouth side cut
Figure 12-12

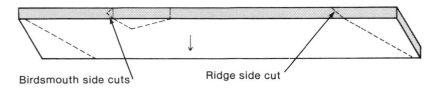

Completed birdsmouth side cuts
Figure 12-13

the plumb line on each side. You will find that this automatically conforms to the side cuts drawn on the bottom of the rafter.

Next cut at the ridge plumb line. Again, if the skill saw table is set at 45°, cutting along the plumb line automatically gives you the proper side cut.

Now you are ready to cut the birdsmouth. With the skill saw table set flat, cut along the seat cut on each side. Then set the skill saw table to 45°, and set the depth of the cut to half the thickness of the rafter. You will be able to cut the birdsmouth plumb cut on one side but not the other because of the way the skill saw is made. After cutting the birdsmouth plumb cut on one side, finish out the birdsmouth with a sharp chisel.

Practical Application: Laying Out Hip, Valley, and Jack Rafters for Equal-Pitch Roofs

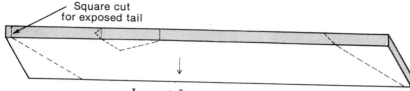

Layout for exposed tail cut
Figure 12-14

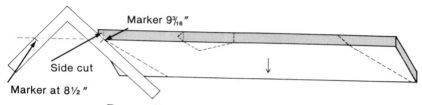

Draw tail side cut to receive facia
Figure 12-15

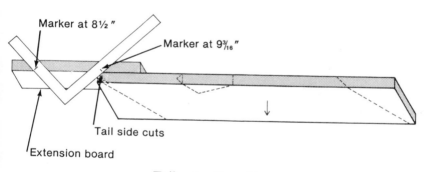

Tail cut, other side
Figure 12-16

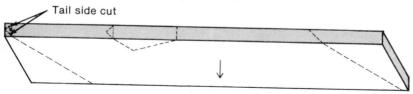

Completed tail cut layout for facia
Figure 12-17

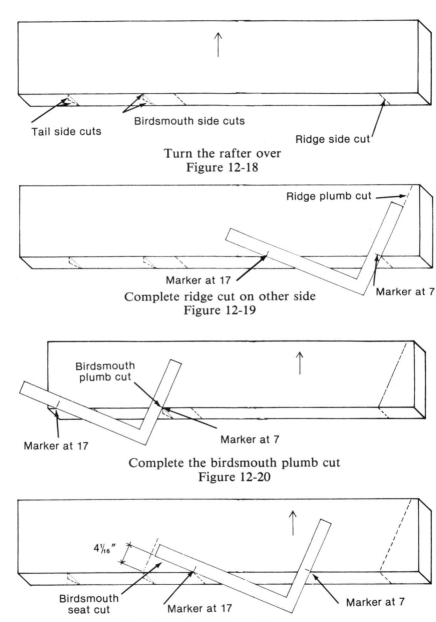

Tail side cuts Birdsmouth side cuts Ridge side cut
Turn the rafter over
Figure 12-18

Ridge plumb cut
Marker at 17
Marker at 7
Complete ridge cut on other side
Figure 12-19

Birdsmouth plumb cut
Marker at 17
Marker at 7
Complete the birdsmouth plumb cut
Figure 12-20

4 1/16"
Birdsmouth seat cut
Marker at 17
Marker at 7
Complete the birdsmouth seat cut
Figure 12-21

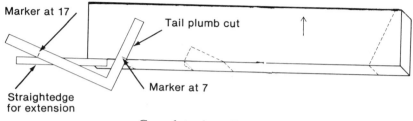

Complete the tail cut
Figure 12-22

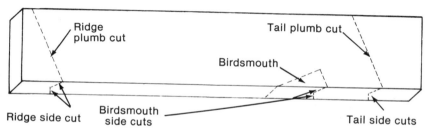

Layout for valley rafter
Figure 12-23

The rafter is now cut and ready to be set.

Laying out a valley rafter involves exactly the same process, except that the tail cut, birdsmouth, and ridge are notched in the opposite direction (Figure 12-23).

When cutting jack rafters, the layout is exactly the same as for common rafters, except for the cut where the jack rafter joins the hip or valley rafter. That cut is made by laying out a plumb line at the proper length (calculated as in Chapter 11) and cutting along that line with the skill saw table set at 45°. The side cut automatically turns out right. The side cut can also be laid out in the same manner that the hip rafter side cut is laid out, but use (unit rise)/12 instead of (unit rise)/17.

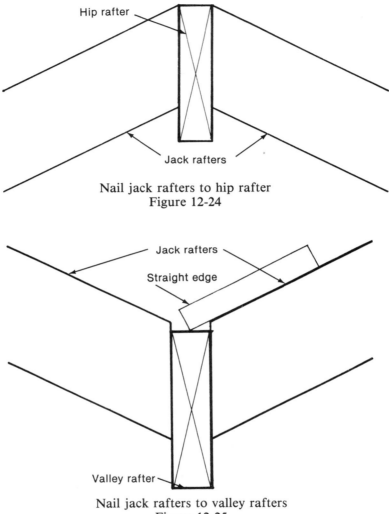

Nail jack rafters to hip rafter
Figure 12-24

Nail jack rafters to valley rafters
Figure 12-25

When setting the jack rafters on a hip roof, nail the jack rafter so the top edge of it lines up with the top edge of the hip rafter (Figure 12-24). When nailing a jack to a valley rafter, lay a straightedge along the top of it, and nail the jack in place when the straightedge lines up with the center line of the valley rafter (Figure 12-25). This is done so the sheathing will lie flat on the valley rafter. When figuring the valley-rafter layout, you dropped the valley rafter to allow for this.

Answers for Chapter Problems

Chapter 2

Problem 1- 6⁷⁄₁₆″
    2- 2³⁄₁₆″
    3- 30¹¹⁄₁₆″
    4- 10⁵⁄₁₆″

Chapter 3

Problem 1- Exact rise = 7″
        Number of treads = 20
        Exact run = 11″
        Tread depth = 11¾″
        First rise = 5⅞″

    2- Exact rise = 5¹⁵⁄₁₆″
        Number of treads = 19
        Exact run = 11¹³⁄₁₆″
        Tread depth = 12¹³⁄₁₆″
        First rise = 7⁷⁄₁₆″

    3- Exact rise = 7⅝″
        Exact run = 11½″
        Tread depth = 12¼″
        First rise = 6⅛″
        Top flight has 7 treads
        Bottom flight has 14 treads
        Landing is 61″ below top floor

Chapter 5

Problem 1- Length of rafter = 172⅛″
        Length of overhang = 31¼″
        Birdsmouth seat cut = 5½″

2- Length of rafter = 319¾"
Length of overhang = 10⅛"
Length between birdsmouth plumb cuts = 170$\frac{7}{16}$"
Birdsmouth seat cuts = 3½"

3- Length of rafter = 207$\frac{9}{16}$"
Length of overhang = 18¼"
Length between birdsmouth plumb cuts = 91¼"
Exterior and intermediate birdsmouth seat cuts = 5⅛"
Ridge birdsmouth seat cut = 2$\frac{9}{16}$"

Chapter 7

Problem 1- 145⅞"
2- 246$\frac{3}{16}$" and 324$\frac{9}{16}$"

Chapter 8

Problem 1- Degree of cut = 14°
Top plate length = 48$\frac{15}{16}$"

Bottom plate layout	Stud length	Top plate layout
0 (1st stud)	91$\frac{7}{16}$"	0
15¼"	95¼"	15¾"
31¼"	99¼"	32$\frac{3}{16}$"
46"	102$\frac{15}{16}$"	47$\frac{7}{16}$"

Problem 2- Degree of cut = 38°
Top plate length = 40⅜″

	Bottom plate layout	Stud length	Top plate layout
West			
	0 (1st stud)	116⅝″	0
	15¼″	128⅜″	19¼″
	30½″	140⅛″	38½″
East			
	0 (1st stud)	116⅝″	0
	16⅝″	129$\frac{7}{16}$″	21″
	30½″	140⅛″	38½″

Chapter 9

Problem 1-

Corner	Distance up or down	Distance right or left	Angle	Distance from point 0
1	-6	0	180°	6′
2	-6	10.167	120°32′48″	11.805′
3	6.25	10.167	58°25′10″	11.934′
4	6.25	2	17°44′41″	6.562′
5	14.25	2	7°59′22″	14.390′
6	14.25	-12	319°53′57″	18.630′
7	-3.75	-12	252°38′46″	12.572′
8	0	-9.5	270°	9.5′

Chapter 11

Problem 1- Hip rafter length = $288\frac{1}{8}''$
Distance to birdsmouth = $249\frac{3}{4}''$
Side cut = $19\frac{1}{4}/17$
Depth of birdsmouth = $4\frac{1}{8}''$
Pitch of hip rafter = $9/17$

2- Valley rafter length = $242\frac{13}{16}''$
Distance to birdsmouth = $212\frac{7}{8}''$
Side cut = $20/17$
Total depth of birdsmouth = $3\frac{5}{16}''$
Pitch of valley rafter = $10\frac{1}{2}/17$

3- Side cut = $13\frac{7}{8}/12$
Jack rafter lengths to short point = $134\frac{3}{8}''$, $97\frac{5}{16}''$, $60\frac{1}{4}''$, $23\frac{3}{16}''$

Appendix B - Multipliers

Roof pitch	Height multipliers	Rake multipliers	Diagonal multipliers
½/12	.0416667	1.0008677	1.4148272
1 /12	.0833333	1.0034662	1.4166667
1½/12	.125	1.0077822	1.4197277
2 /12	.1666667	1.0137938	1.4240006
2½/12	.2083333	1.0214709	1.4294764
3 /12	.25	1.0307764	1.4361407
3½/12	.2916667	1.0416667	1.443977
4 /12	.3333333	1.0540926	1.4529663
4½/12	.375	1.0680005	1.4630875
5 /12	.4166667	1.0833333	1.4743172
5½/12	.4583333	1.1000316	1.4866302
6 /12	.5	1.118034	1.5
6½/12	.5416667	1.1372787	1.5143985
7 /12	.5833333	1.1577037	1.5297966
7½/12	.625	1.1792476	1.5461646
8 /12	.6666667	1.2018504	1.5634719
8½/12	.7083333	1.2254534	1.5816877
9 /12	.75	1.25	1.6007811
9½/12	.7916667	1.2754357	1.6207209
10 /12	.8333333	1.3017083	1.6414763
10½/12	.875	1.3287682	1.6630168
11 /12	.9166667	1.3565684	1.6853124
11½/12	.9583333	1.3850642	1.7083333
12 /12	1.0	1.4142136	1.7320508
12½/12	1.0416667	1.443977	1.7564366
13 /12	1.0833333	1.4743172	1.7814632
13½/12	1.125	1.5051993	1.807104
14 /12	1.1666667	1.5365907	1.8333333
14½/12	1.2083333	1.5684609	1.8601262
15 /12	1.25	1.6007811	1.8874586
15½/12	1.2916667	1.6335246	1.9153075
16 /12	1.3333333	1.6666667	1.9436506
16½/12	1.375	1.7001838	1.9724667
17 /12	1.4166667	1.7340543	2.0017354
17½/12	1.4583333	1.7682579	2.031437
18 /12	1.5	1.8027756	2.0615528

Appendix C
Fraction to Decimal Conversions

$1/16$	=	.0625	$9/16$	=	.5625
$1/8$	=	.125	$5/8$	=	.625
$3/16$	=	.1875	$11/16$	=	.6875
$1/4$	=	.25	$3/4$	=	.75
$5/16$	=	.3125	$13/16$	=	.8125
$3/8$	=	.375	$7/8$	=	.875
$7/16$	=	.4375	$15/16$	=	.9375
$1/2$	=	.5	1	=	1.0

Appendix D
Decimal to Fraction Conversions

0.0 to .03124	=	0
.03125 to .09374	=	$1/16$
.09375 to .15624	=	$1/8$
.15625 to .21874	=	$3/16$
.21875 to .28124	=	$1/4$
.28125 to .34374	=	$5/16$
.34375 to .40624	=	$3/8$
.40625 to .46874	=	$7/16$
.46875 to .53124	=	$1/2$
.53125 to .59374	=	$9/16$
.59375 to .65624	=	$5/8$
.65625 to .71874	=	$11/16$
.71875 to .78124	=	$3/4$
.78125 to .84374	=	$13/16$
.84375 to .90624	=	$7/8$
.90625 to .96874	=	$15/16$
.96875 to 1.03124	=	1

Index

A

Added height 105-106
Algebraic logic 6
Answers 211-214

B

Balusters 9, 13
Batterboards 148
Beam
   bearing 67
   glulam 78, 108-112
Bearing (rafter)
   full 67, 69, 105
   partial 67-69
Bearing wall 67
Bearing-wall heights . 103-116
Birdsmouth cuts
   common rafters . 64, 66-69,
                  71, 100-101
   hip rafters 174-175,
             201-202, 204-208
   valley rafters 184, 209
Birdsmouths
   length between . . 66, 68-71
   run between 64, 66-71,
                     73, 75
Board-and-batten siding . . 9,
                  10-11, 21-25
Bottom plate
   layout 120-122
   length 120

C

Calculator
   color display 7
   memory 118
   programming 118
Calculators
   algebraic 6
   programmable 118
   reverse Polish logic 6
   sequential 6
   Texas Instruments 8
Cant strip 134-136
Carriage, stair 57-60

Center-to-center dimension 10
Common difference . 188-189
Common rafters
  cutting 99-101
  fitting 101
  layout 63-72
Control point (foundation) .. 149-150
Corner stakes 148
Cosine 170, 171
Course
  definition 9
  dimension 15
  largest acceptable 9-13
Courses
  equally spaced 9-37
  number of 15
Cripple jack rafter 63
Crowning
  rafters 99-100, 201-202, 205
  stair carriages 57

D

Decimal feet 150
Decimal fractions 14, 216
Decking, calculating courses 9, 11, 13, 26
Denominator 14
Desired rise 38, 39, 41
Desired run 38, 41
Diagonal multiplier 176
Difference, common . 188-189
Drop, valley rafter .. 184-185, 210

E

Equally spaced courses .. 9-37
Exact rise 38-39, 41-42
Exact run 38-39, 41-42

F

Facia 174-175, 205, 207
First rise, stair 38, 41, 43
First stud length 119-120, 123
Footing ... 153, 154, 160, 172
Foundation layout 7, 148-152

Fractions, converting . 14, 216
Framing square 5, 55-56, 99-101, 200-202
Full bearing, rafter 67, 69, 105
Functions (mathematical) .. 6

G

Gable roof 65, 67, 72, 76
Gable wall 129-130, 134
Glulam purlin beams .. 78-79, 84

H

Handrails, spacing .. 9, 32, 33
Height
  bearing-wall 103-116
  stair 38-40
Height, added 105-106
Height multiplier 105
Hip rafters
  calculating 173-177
  laying out 200-208
Hip roof 173-174

I

Inverse tangent 146

J

Jack rafter
  calculating ... 173, 187-190
  laying out 209-210
Joists, calculating courses ... 9-10, 19-20

L

Landing, stair .. 38, 47, 50-52
Largest acceptable course ... 9-13
Layout
  bottom plate 120-122
  common rafter 63-72
  equally spaced courses 9-37

foundation 148-152, 167-172
hip rafter 200-208
jack rafter 209-210
picket fence 31
stair 38, 55
top plate . 117-119, 123-125
valley rafter 201-210
Length
  between birdsmouths .. 66, 68-71
  bottom plate 120
  stair 38-39
  top plate 117, 122-123
Level cuts.............. 201
Logic, calculator.......... 6

M

Mathematics 5
Memory, calculator 118
Multipliers
  diagonal 176-177, 215
  height........... 105, 215
  rake 123, 215

N

Nonbearing partition 114
Nosing, stair 38, 41, 43
Numerator.............. 14

O

Overhang
  rafter, length 66, 69-71
  rafter, run.... 64-66, 69-71
  shingle 12, 29

P

Paneling, spacing .. 9, 13, 17, 34-35
Partial bearing, rafter.. 67-69
Partition, nonbearing ... 114
Pickets, equally spaced-courses.... 9, 13, 15, 31-32

Piers 155, 156
Pitch, roof ... 84, 89, 103-105
Plumb bob 168-169
Plumb cut, birdsmouth
  common rafter .. 64, 66-68, 100-101, 102
  hip rafters 174-177, 201-208
  valley rafters 183-184
Point chart 150-151
Point person 168-170
Polar coordinates 149
Programmable calculator 118
Purlin beams 78

R

Rafter
  bearing 67-69, 105
  cutting.... 99-101, 205-209
  fitting............... 101
  layout 63-72, 200-210
  length............. 66, 68
  run 64, 65-67, 70
Rafter square 5, 55-56, 99-101, 200-202
Rafters
  common 63-72, 99-102
  cripple jack 63
  hip 174-177, 200-208
  jack . 173, 187-190, 209-210
  valley 183-187, 209
Rake multiplier...... 69, 215
Rake-wall layout .. 7, 117-125
Reference points 171
Reverse Polish logic 6
Ridge
  cut .. 174-175, 204-205, 208
  line 63
  pole 63, 67
  purlin 78
Ridgecap course 12, 29
Rise
  desired 38, 39, 41
  exact 41-42
  first 38, 41, 43
  unit 65
Roof pitch ... 84, 89, 103-105
Roofs
  gable 65, 67, 72, 76

hip 173-174
shed 65, 83
Run
  between birdsmouths .. 64,
    66-71, 73, 75
  desired 38, 41
  exact 38-39, 41-42
  of overhang .. 64-66, 69-71
  of rafter 64, 65-67, 70
  unit 104, 105, 120, 174

S

Seat cut .. 64, 66-68, 100-102,
    105, 202
Sequential logic 6
Shakes, exposure 9, 12
Shed roof 65, 83
Shingle overhang ... 9, 12, 29
Shiplap siding 13, 34
Short point length, studs . 118
Side cuts (hip rafters) ... 174,
    176-177, 201-208
Siding, equally spaced
  courses 9, 13, 21
Slope distance 170-172
Sloping top plate 118
Stair
  carriages 57-58, 60-61
  height 38-40
  layout 38, 55
  length 38-39
Stud length (rake wall) .. 117,
    118-120, 123-135
Subtread thickness 39, 42-43,
    60
Support run 105-106
Symmetrical layout . 9, 13, 34

T

Tail cut (rafter) 100-101,
    202-204, 207, 209
Tangent, inverse 120
Tongue-and-groove siding 13,
    34
Top plate
  layout ... 117-119, 123-125
  length 117, 122-123
Transit 148-149, 164, 168-172
Tread depth 38, 42-43

U

Unequal pitch roof 89,
    139-140
Unit rise 65, 105
Unit run .. 104, 105, 120, 174

V

Valley rafters
  cutting 209
  dropping 184-185
  laying out 183-185

W

Walls
  bearing 67, 103-106
  gable 129-130, 134
  rake 117-125

Other Practical References

Drywall Contracting
How to do professional quality drywall work, how to plan and estimate each job, and how to start and keep your drywall business thriving. Covers the eight essential steps in making any drywall estimate, how to achieve the six most commonly-used surface treatments, how to work with metal studs, and how to solve and prevent most common drywall problems. **288 pages, 5½ x 8½, $18.25**

Carpentry for Residential Construction
How to do professional quality carpentry work in homes and apartments. Illustrated instructions show you everything from setting batter boards to framing floors and walls, installing floor, wall and roof sheathing, and applying roofing. Covers finish carpentry, also: How to install each type of cornice, frieze, lookout, ledger, fascia and soffit; how to hang windows and doors; how to install siding, drywall and trim. Each job description includes the tools and materials needed, the estimated manhours required, and a step-by-step guide to each part of the task. **400 pages, 5½ x 8½, $19.75**

Carpentry in Commercial Construction
Covers forming, framing, exteriors, interior finish and cabinet installation in commercial buildings: designing and building concrete forms, selecting lumber dimensions, grades and species for the design load, what you should know when installing materials selected for their fire rating or sound transmission characteristics, and how to plan and organize the job to improve production. Loaded with illustrations, tables, charts and diagrams. **272 pages, 5½ x 8½, $19.00**

Roof Framing
Frame any type of roof in common use today, even if you've never framed a roof before. Shows how to use a pocket calculator to figure any common, hip, valley, and jack rafter length in seconds. Over 400 illustrations take you through every measurement and every cut on each type of roof: gable, hip, Dutch, Tudor, gambrel, shed, gazebo and more. **480 pages, 5½ x 8½, $19.50**

Rough Carpentry
All rough carpentry is covered in detail: sills, girders, columns, joists, sheathing, ceiling, roof and wall framing, roof trusses, dormers, bay windows, furring and grounds, stairs and insulation. Many of the 24 chapters explain practical code approved methods for saving lumber and time without sacrificing quality. Chapters on columns, headers, rafters, joists and girders show how to use simple engineering principles to select the right lumber dimension for whatever species and grade you are using. **288 pages, 8½ x 11, $16.00**

National Construction Estimator
Current building costs in dollars and cents for residential, commercial and industrial construction. Prices for every commonly used building material, and the proper labor cost associated with installation of the material. Everything figured out to give you the "in place" cost in seconds. Many time-saving rules of thumb, waste and coverage factors and estimating tables are included. **528 pages, 8½ x 11, $18.50. Revised annually.**

Building Cost Manual
Square foot costs for residential, commercial, industrial, and farm buildings. In a few minutes you work up a reliable budget estimate based on the actual materials and design features, area, shape, wall height, number of floors and support requirements. Most important, you include all the important variables that can make any building unique from a cost standpoint. **240 pages, 8½ x 11, $14.00. Revised annually**

Building Layout
Shows how to use a transit to locate the building on the lot correctly, plan proper grades with minimum excavation, find utility lines and easements, establish correct elevations, lay out accurate foundations and set correct floor heights. Explains planning sewer connections, leveling a foundation out of level, using a story pole and batterboards, working on steep sites, and minimizing excavation costs. **240 pages, 5½ x 8½, $11.75**

Builder's Guide to Accounting Revised
Step-by-step, easy to follow guidelines for setting up and maintaining an efficient record keeping system for your building business. Not a book of theory, this practical, newly-revised guide to all accounting methods shows how to meet state and federal accounting requirements, including new depreciation rules, and explains what the tax reform act of 1986 can mean to your business. Full of charts, diagrams, blank forms, simple directions and examples. **304 pages, 8½ x 11, $17.25**

Computers: The Builder's New Tool
Shows how to avoid costly mistakes and find the right computer system for your needs. Takes you step-by-step through each important decision, from selecting the software to getting your equipment set up and operating. Filled with examples, checklists and illustrations, including case histories describing experiences other contractors have had. If you're thinking about putting a computer in your construction office, you should read this book before buying anything. **192 pages, 8½ x 11, $17.75**

Handbook of Construction Contracting Vol. 1 & 2
Volume 1: Everything you need to know to start and run your construction business; the pros and cons of each type of contracting, the records you'll need to keep, and how to read and understand house plans and specs to find any problems before the actual work begins. All aspects of construction are covered in detail, including all-weather wood foundations, practical math for the jobsite, and elementary surveying. **416 pages, 8½ x 11, $21.75**

Volume 2: Everything you need to know to keep your construction business profitable; different methods of estimating, keeping and controlling costs, estimating excavation, concrete, masonry, rough carpentry, roof covering, insulation, doors and windows, exterior finish, specialty finishes, scheduling work flow, managing workers, advertising and sales, spec building and land development and selecting the best legal structure for your business. **320 pages, 8½ x 11, $24.75**

Carpentry Estimating
Simple, clear instructions show you how to take off quantities and figure costs for all rough and finish carpentry. Shows how much overhead and profit to include, how to convert piece prices to MBF prices or linear foot prices, and how to use the tables included to quickly estimate manhours. All carpentry is covered: floor joists, exterior and interior walls and finishes, ceiling joists and rafters, stairs, trim, windows, doors, and much more. Includes sample forms, checklists, and the author's factor worksheets to save you time and help prevent errors. **320 pages, 8½ x 11, $25.50**

E-Z Square
This round, plastic computer is designed to help you save time in calculating diagonal dimensions in laying out excavations, footings, foundations, forms, walls, and plates. It also computes brick and block quantities for any known wall area and cubic yardage for excavating. It reads like a carpenters rule and is an invaluable time saver in layout work. **$9.25 each.**

Spec Builder's Guide
Explains how to plan and build a home, control your construction costs, and then sell the house at a price that earns a decent return on the time and money you've invested. Includes professional tips to ensure success as a spec builder: how government statistics help you judge the housing market, cutting costs at every opportunity without sacrificing quality, and taking advantage of construction cycles. Every chapter includes checklists, diagrams, charts, figures, and estimating tables. **448 pages, 8½ x 11, $24.00**

Stair Builders Handbook
If you know the floor to floor rise, this handbook will give you everything else: the number and dimension of treads and risers, the total run, the correct well hole opening, the angle of incline, the quantity of materials and settings for your framing square for over 3,500 code approved rise and run combinations—several for every 1/8 inch interval from a 3 foot to a 12 foot floor to floor rise. **416 pages, 8½ x 5½, $13.75**

Rafter Length Manual
Complete rafter length tables and the "how to" of roof framing. Shows how to use the tables to find the actual length of common, hip, valley and jack rafters. Shows how to measure, mark, cut and erect the rafters, find the drop of the hip, shorten jack rafters, mark the ridge and much more. Has the tables, explanations and illustrations every professional roof framer needs. **369 pages, 8½ x 5½, $12.25**

Finish Carpentry
The time-saving methods and proven shortcuts you need to do first class finish work on any job: cornices and rakes, gutters and downspouts, wood shingle roofing, asphalt, asbestos and built-up roofing, prefabricated windows, door bucks and frames, door trim, siding, wallboard, lath and plaster, stairs and railings, cabinets, joinery, and wood flooring. **192 pages, 8½ x 11, $10.50**

Dial-A-Length Rafterule
Just set the dial and read off the answers. The Dial-A-Length Rafterule determines the hip, valley, common, and jack rafter length for any span with "saw cut" accuracy. Gives plumb, level, and side cut data plus angle in degrees and minutes for 21 pitch settings. Scales graduated like any carpenters rule. Circular design gives accuracy to $\frac{1}{16}$". The Dial-A-Length Rafter rule produces nothing but **correct** answers. **$9.50 each**

Remodeler's Handbook
The complete manual of home improvement contracting: Planning the job, estimating costs, doing the work, running your company and making profits. Pages of sample forms, contracts, documents, clear illustrations and examples. Chapters on evaluating the work, rehabilitation, kitchens, bathrooms, adding living area, re-flooring, re-siding, re-roofing, replacing windows and doors, installing new wall and ceiling cover, re-painting, upgrading insulation, combating moisture damage, estimating, selling your services, and bookkeeping for remodelers. **416 pages, 8½ x 11, $18.50**

Visual Stairule
Easy, simple settings on this pocket-sized adjustable blueprint indicate proper number and height of risers and various combinations, proper number and widths of treads for combination selected, extent of total run, head room dimensions, stringer length, well opening. An invaluable tool for contractors and carpenters who want to take the problems out of stair planning. **$7.50 each.**

Manual of Professional Remodeling
This is the practical manual of professional remodeling written by an experienced and successful remodeling contractor. Shows how to evaluate a job and avoid 30-minute jobs that take all day, what to fix and what to leave alone, and what to watch for in dealing with subcontractors. Includes chapters on calculating space requirements, repairing structural defects, remodeling kitchens, baths, walls and ceilings, doors and windows, floors, roofs, installing fireplaces and chimneys (including built-ins), skylights, and exterior siding. Includes blank forms, checklists, sample contracts, and proposals you can copy and use. **400 pages, 8½ x 11, $18.75**

Video: Roof Framing 1
A complete step-by step training video on the basics of roof cutting by Marshall Gross, the author of the book **Roof Framing**. Shows and explains calculating rise, run, and pitch, and laying out and cutting common rafters. **90 minutes, VHS, $80.00**

Video: Roof Framing 2
A complete training video on the more advanced techniques of roof framing by Marshall Gross, the author of **Roof Framing,** shows and explains layout and framing an irregular roof, and making tie-ins to an existing roof. **90 minutes, VHS, $80.00**

Craftsman Book Company
6058 Corte del Cedro
P. O. Box 6500
Carlsbad, CA 92008

10 Day Money Back
GUARANTEE

Phone Orders

In a Hurry?
We accept phone orders charged to your MasterCard or Visa.
Call (619) 438-7278

Mail Orders
We pay shipping when your check covers your order in full.

- ☐ 17.25 Builder's Guide to Accounting
- ☐ 14.00 Building Cost Manual
- ☐ 11.75 Building Layout
- ☐ 25.50 Carpentry Estimating
- ☐ 19.75 Carpentry for Residential Construction
- ☐ 19.00 Carpentry in Commercial Construction
- ☐ 17.75 Computers: The Builder's New Tool
- ☐ 9.50 Dial-A-Length Rafterule
- ☐ 18.25 Drywall Contracting
- ☐ 9.25 E-Z Square
- ☐ 10.50 Finish Carpentry
- ☐ 21.75 Handbook of Construction Contracting Vol. 1
- ☐ 24.75 Handbook of Construction Contracting Vol. 2
- ☐ 18.75 Manual of Professional Remodeling
- ☐ 18.50 National Construction Estimator
- ☐ 12.25 Rafter Length Manual
- ☐ 18.50 Remodeler's Handbook
- ☐ 19.50 Roof Framing
- ☐ 16.00 Rough Carpentry
- ☐ 24.00 Spec Builder's Guide
- ☐ 13.75 Stair Builder's Handbook
- ☐ 80.00 Video: Roof Framing 1
- ☐ 80.00 Video: Roof Framing 2
- ☐ 7.50 Visual Stairule
- ☐ 16.25 Carpentry Layout

Name (Please print clearly) _____

Company _____

Address _____

City _____ **State** _____ **Zip** _____

Send check or money order
Total enclosed _____ (In California add 6% tax)
If you prefer use your ☐ Visa ☐ MasterCard
Card no. _____
Expiration date _____ Initials _____